Longman

Foundation Science 1

for GCSE

Mark Levesley

Jackie Hardie

Richard O'Regan

Sarah Pitt

Nicky Thomas

Bob Wakefield

Contents for Foundation Science

What is life?

What makes something alive?

Life process	What it means	Life process	What it means
Movement 	Living things can move all or part of themselves.	**Respiration** 	Living things get energy from food (often by using oxygen).
Reproduction 	Living things can produce offspring.	**Excretion** 	Living things get rid of the waste substances they produce.
Sensitivity 	Living things can sense things in their surroundings.	**Nutrition** 	Living things get food. Plants use energy from light to help them make food. Animals eat plants or other animals.
Growth 	Living things can increase the size of their bodies by adding mass and (often) new cells.		

 The seven life processes.

We all know what it feels like to be alive, but it is hard to explain what life is.

Living things are called **organisms** and include all the plants and animals on earth. Every living thing does all seven of the life processes in table A.

 1 List the seven life processes.

2 What is an organism?

Living things have different life spans. An adult mayfly lives for only one day. Humans are alive for about 80 years. An oak tree may have a life span of over 200 years. One bristle cone pine in Arizona, USA, is believed to be 4000 years old!

All living things are made of **cells**. Cells are very small. About 40 cells from the inside of your cheek would fit into a line 1 mm long.

3 What is your body made from?

Each part of the cell has a different job. Even though cells can look different, they have some features which are the same.

4 What separates a cell from its surroundings?

5 What controls the cell?

6 Where do the chemical reactions happen in a cell?

Summary

All living things are called _____. They all show the _____ life processes. All living things are made from _____. All cells have some _____ that are the same. The nucleus _____ the cell. The cell _____ separates the cell from its surroundings, and controls the substances which go in and out of a cell. _____ reactions take place in the _____.

cells   chemical   controls
cytoplasm   features
membrane   organisms   seven

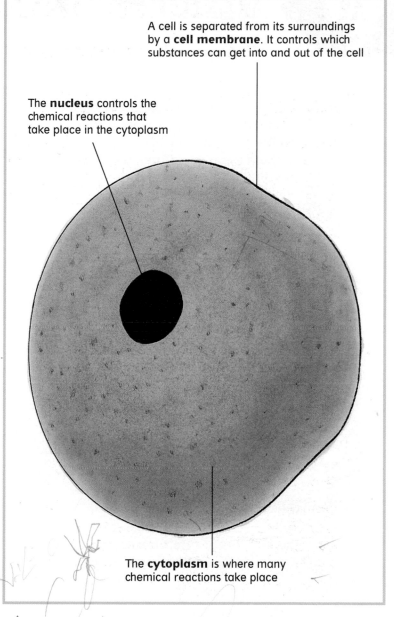

The **nucleus** controls the chemical reactions that take place in the cytoplasm

A cell is separated from its surroundings by a **cell membrane**. It controls which substances can get into and out of the cell

The **cytoplasm** is where many chemical reactions take place

B The parts of an animal cell.

7 A friend tells you that their car is alive.
  a) Give 3 reasons why your friend is wrong.
  b) Give 3 ways in which a car is similar to something that is alive.

8 Write a short sentence to explain the main job of each of these parts of a cell:
  a) nucleus
  b) cell membrane
  c) cytoplasm.

Different kinds of cell

How are cells adapted to do their jobs?

A motorbike is made up of many different parts, so is a house, a computer and so are you! Cells from different parts of your body come in different shapes and sizes. They all have jobs to do and their shapes help them to do their jobs. We say that the cells are **adapted** to do their jobs (or **functions**).

Sperm cells have a 'tail' which moves from side to side helping the sperm cell to swim to the egg cell.

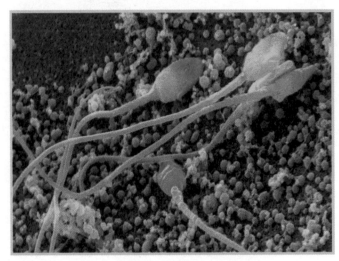

Your **white blood cells** can move by changing their shape, so they can surround bacteria.

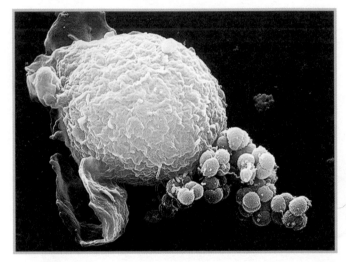

Nerve cells have lots of fine threads on their surface so they can make connections with lots of other nerve cells. This means they can pass electrical messages (called **impulses**) from one cell to another.

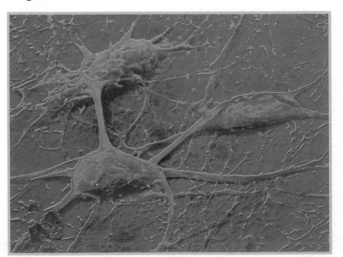

Muscle cells are long and thin. These cells can **contract** (get shorter). When groups of muscle cells contract they move part of your body.

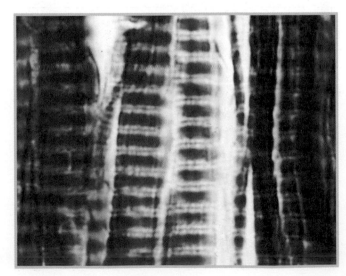

Some cells in your lungs, called **ciliated epithelial cells**, have **cilia**, which look like small hairs. The cilia can move to sweep dirt out of your lungs.

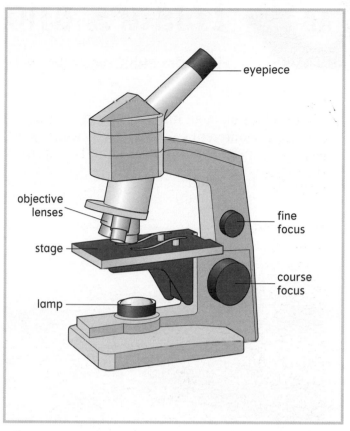

F *A microscope.*

Labels on microscope: eyepiece, objective lenses, stage, lamp, fine focus, course focus

?

1 Why do cells have different shapes?

2 How are sperm cells adapted to their job?

3 How do white blood cells move?

P

You can look at some human cells with a microscope. How could you measure the size of cells with a microscope?

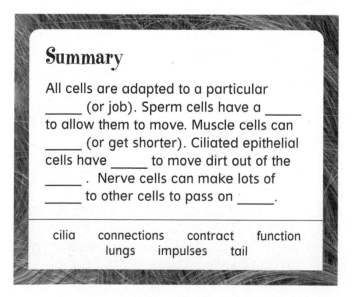

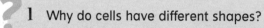

Summary

All cells are adapted to a particular _____ (or job). Sperm cells have a _____ to allow them to move. Muscle cells can _____ (or get shorter). Ciliated epithelial cells have _____ to move dirt out of the _____ . Nerve cells can make lots of _____ to other cells to pass on _____.

cilia	connections	contract	function
	lungs	impulses	tail

Looking at cells

You need to use a microscope to see cells clearly. When you look down a microscope, you see magnified cells. The cells look bigger than they are and you can see the nucleus, the cytoplasm and the cell membrane.

?

4 What is a microscope used for?

5 How is a white blood cell suited to its job of killing bacteria?

6 Find out the function of red blood cells, and how they are adapted to their function.

Tissues and organs

How do cells work together?

Your body is made of millions of cells. The cells in your body have many different shapes and functions (jobs). All the cells in your body need to work together to keep you alive and healthy.

Cells are often grouped together with cells of the same type to do one job. A group of cells like this is called a **tissue**.

? 1 What is a tissue?

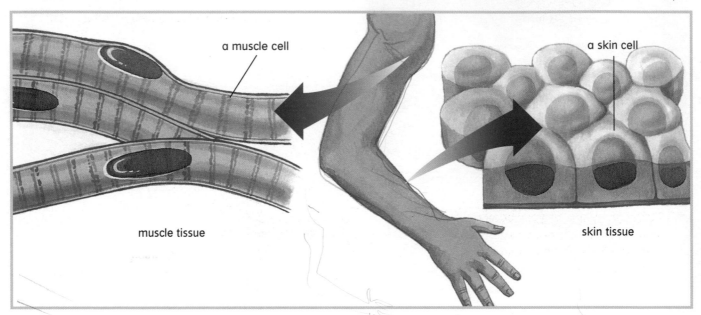

a muscle cell

muscle tissue

a skin cell

skin tissue

A

Muscle cells are grouped into **muscle tissue**. When all the cells **contract** (get shorter) at the same time, they make the muscle tissue shorter. Muscle tissue moves parts of your body.

Other cells are grouped to cover the surfaces of your body. These covering tissues are called **epithelium** tissue.

? 2 Write down the name of one part of the body that might contain epithelium tissue.

Some cells are grouped together to form **glandular** tissue. Gland cells work to make fluids called **secretions**, which leave the cells and go to other parts of the body. For example, in your eyes the cells of your tear glands make the secretions we call tears. These tears pass out from the gland cells and are used to clean the surface of your eye.

B

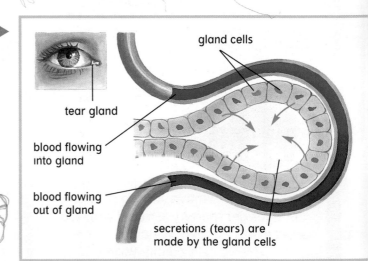

gland cells

tear gland

blood flowing into gland

blood flowing out of gland

secretions (tears) are made by the gland cells

? 3 Which tissue helps you move your body?

4 a) What does glandular tissue produce?
b) Write down the name of one gland.

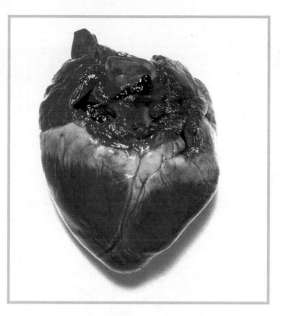

A group of different tissues all working together is called an **organ**. Your eyes, stomach and heart are all organs. Each organ has one main job. Your stomach is made of muscle, nerve and glandular tissue. Your stomach's main job is to mix the food you swallow with secretions from its glandular tissue. The muscle tissue in the stomach walls contracts and so the food inside it is mixed with the secretions.

Organs often work together in **organ systems**. Your mouth, gullet, stomach, intestines, pancreas and liver are all organs that are found in the **digestive system** which breaks down the food that you swallow.

The organs in the digestive system work together.

C The heart is an organ made from muscle and nerve tissue.

5 Write down the names of three tissues found in the stomach.

P Draw an outline of your body on a large piece of paper. Draw in the places where you find your brain, lungs, liver and kidneys.

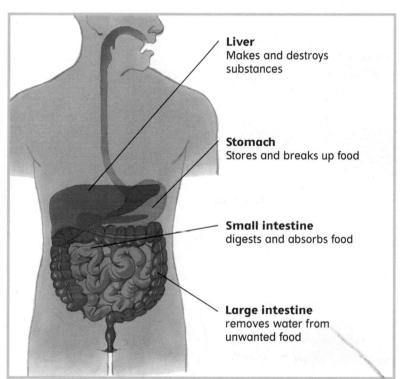

Liver
Makes and destroys substances

Stomach
Stores and breaks up food

Small intestine
digests and absorbs food

Large intestine
removes water from unwanted food

Summary

A _____ contains cells of the same type working together. The tissue that helps to move the body is _____ tissue. In your eyes, _____ cells are grouped in tear glands. These glands make _____ called tears. An _____ contains several tissues working together. Organs may work together to make an organ _____. The _____ system breaks down the food that you swallow.

digestive   gland   muscle   organ
    secretions   system   tissue

6 Write down whether each of the following is a cell, a tissue or an organ:

a) sperm
b) a gland
c) the eye
d) the stomach
e) the brain.

7 What is the job of the stomach?

8 What is the difference between a tissue and an organ?

9 Explain the function of tear glands in the eye.

11

A4 Food

Why do we need to eat?

Every living organism needs energy and chemicals to build new cells. The energy and the chemicals come from food. **Nutrition** means eating food to keep us healthy and to keep our cells alive. We eat food that comes from animals and from plants. What you eat is called your **diet**.

You need food to:

- give your body energy
- supply materials that can be used for growth and repairing your body
- supply materials that are needed to keep you healthy.

1 Why do we eat plants and animals?

2 Write down three things your body needs food for.

All foods contain water and other chemicals called **nutrients**. These nutrients can be divided into 5 groups: **proteins, carbohydrates, fats, vitamins** and **minerals**. To stay healthy you need to eat the right amounts of these nutrients. You also need to eat foods containing **fibre**, and you also need **water**.

3 **a)** List the 5 kinds of nutrients that your body needs.
b) What other substances does your body need?

About 75% of your body is water. It is used to dissolve other substances and carry them around your body. Water also helps you get rid of waste chemicals.

Most of every cell in your body is made of **protein**. Protein is needed to keep your body growing and to help replace cells that have worn out. Protein is also needed for repairing cells that are damaged.

Proteins are giant molecules made up of long chains of smaller molecules. These smaller molecules are called **amino acids**.

4 Look at diagram B. Which foods have protein in them?

A *People in different countries eat different types of food.*

B *Different foods contain different amounts of nutrients.*

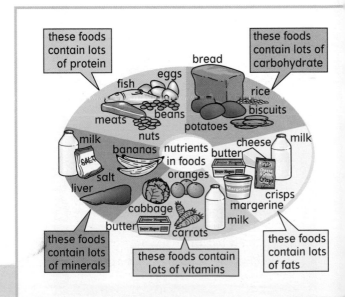

Carbohydrates are needed for energy. **Sugars** arc carbohydrates, and so is **starch**.

Sucrose is one form of sugar. It is found in fruit juices and in the stems of plants like sugar cane. Another sugar is **glucose**. Sugars dissolve in water and taste sweet. Starch is found in some plant leaves, potatoes, and rice. Starch is made of a long chain of glucose molecules. It does not taste sweet.

5 Which foods contain the carbohydrate known as starch?

6 Which foods contain the carbohydrate known as glucose?

Fats can be liquid, like olive oil, or solid like the fat on meat. Fats do not dissolve in water. There is a fat layer underneath your skin which helps to keep your body warm. Fat around delicate organs, like your kidneys, protects them from damage. Fat is needed to make cell membranes and can also be used as an energy store.

7 Look at diagram B. Write down four foods which contain a lot of fat.

Minerals and **vitamins** are needed in very small quantities to help the body to work properly.

The cell walls of plants cells are made of a sunstance called **cellulose**. Cellulose is also called **fibre**. Cellulose cannot be digested by your body, but it gives the muscles of your gut something to squeeze. This means your gut walls get exercise and keep fit!

8 What is fibre?

9 Why does your body need fat?

10 a) Explain why your body needs vitamins and minerals.
b) Find out about the diseases scurvy and rickets.

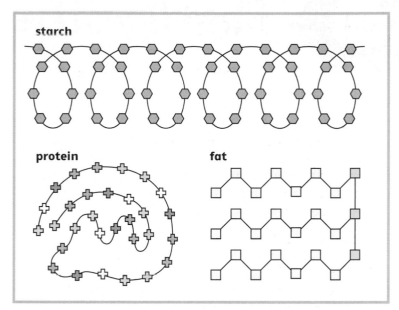

starch

protein

fat

 C *Different food substances have different structures.*

P How could you find out if foods contain starch or protein?

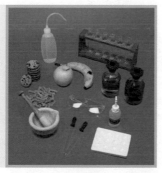

D

Summary

Proteins, carbohydrates, fats, vitamins and minerals are called _N___ . Proteins are chains of amino _A___ and they are used for _Grth_. Starch and _sr_ are types of carbohydrate. Carbohydrates are used for _En__. Fats can be used to make cell _Mbrnes_ Fats can also be used to help keep us _wrth_ and for energy. Minerals and _vit_ are needed in small amounts to keep the body healthy. About 75% of your body is _wtr_ that is used to _duk_ substances.

acids   dissolve   energy   growth
membranes   nutrients   sugars
vitamins   warm   water

The digestive system

What happens to food as it travels down your gut?

Your gut is all the tubes in your digestive system. It is not a straight tube. It is coiled up so that the 6·5 metres of your gut can be fitted inside your body. Your gut is made of several tissues, including muscular and glandular tissues. The nutrients in your food must dissolve so that they can move through the wall of the small intestine and into your blood. Proteins, starch and fats do not dissolve. They have to be split into smaller units which do dissolve. The splitting of the molecules is called **digestion**.

1 What is digestion?

Inside the gut, food is mixed with **enzymes** and other substances to help digest the food. Some of these substances are produced by organs like the liver and pancreas. Your gut and all the organs needed to digest your food are called the **digestive system**. Useful products are **absorbed** through the gut wall and get into your blood.

! You can swallow food when you are upside down. Astronauts can swallow their food because muscles in the gullet push the food to their stomach, whatever position they are in!

Digestion begins in your mouth. When you chew food you chop it into smaller pieces. The food is mixed with **saliva**, which contains **enzymes** and also a slimy substance called **mucus** which helps food slip down your **gullet** (or throat). Your tongue shapes the food into a small ball that is easy to swallow.

2 How is food changed in your mouth?

The ball of food is pushed down the gullet until it reaches the **stomach**.

Your stomach is a bag with lots of muscle in its walls. There are rings of muscle at each end of the stomach which work like elastic bands, so food can be held in your stomach for a couple of hours.

Your stomach muscles work to squeeze on the food. This mixes the food with a **digestive juice** made by the glandular tissue in the stomach lining. This **secretion** is called **gastric juice**. After several hours, your meal becomes a soup-like liquid. The liquid is squirted into the next part of your gut, the small intestines.

3 a) How long is food kept in the stomach?
b) What keeps the food in the stomach?

4 What do the stomach muscles do?

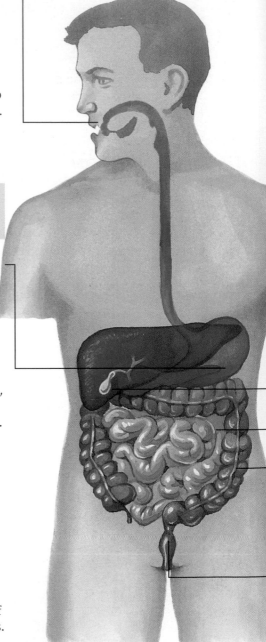

In the **small intestine**, digestive juices are secreted by the glandular tissue in your **pancreas** and the wall of the small intestine. **Bile** is also added. Bile is made in the **liver** and stored in the **gall bladder** until it is needed.

After digestion, the dissolved nutrients move across the wall of the small intestine and into the blood. This is called **absorption**.

Some parts of your food cannot be digested and absorbed. The cell walls of plants are made of cellulose, which you cannot digest. The cellulose (called **fibre**), makes the food bulky, and gives the muscles something to squeeze against as the food continues its journey through the digestive system.

5 Name two substances that are added to the food in the small intestine.

The undigested remains of your meal move into the **large intestine**, where water is absorbed into the blood.

In the last part, called the **rectum**, the remains are formed into solid waste (**faeces**) and stored. When you go to the toilet they are passed out through the **anus**.

6 What is absorbed in the large intestine?

P If you add a blue chemical called DCPIP to fruit juice, the vitamin C in the fruit juice turns the DCPIP colourless. How would you find out which fruit juices contained the most vitamin C?

7 What does the liver make that helps digestion?

8 Copy and complete table B to show what happens to food as it passes down your gut.

Part of gut	What happens to food

Summary

Digestion starts in the _____, where food is chopped up and mixed with _____. It then moves down the _____ to the stomach, where more _____ _____ are added. Food stays in the _____ for a couple of hours and then moves to the _____ intestine. _____ and more digestive juices are added here. Water is absorbed in the _____ intestine. Undigested food, such as _____, is formed into faeces in the _____ and passed out of the _____.

anus   bile   digestive juices   fibre   gullet
large   mouth   rectum   saliva   small   stomach

Breaking down food

What substances help digestion?

The food that goes into your mouth is changed as it goes through you. The journey through your body takes between 24 and 48 hours.

1 Look at picture A. List all the organs in your digestive system.

2 How long does food take to go from your mouth to your anus?

Digestive juices and other substances are added to your food as it goes through your digestive system. These substances all have particular jobs to do to break up your food into smaller particles so that nutrients can be absorbed. Digestive juices contain **enzymes**. These are chemicals that speed up the breaking apart of large molecules in your food.

3 List three organs that produce amylase enzymes.

4 List three organs that produce protease enzymes.

5 List two organs that produce lipase enzymes.

Each kind of enzyme digests one type of nutrient:

- carbohydrase enzymes digest starch
- protease enzymes digest protein
- lipase enzymes digest fats.

6 Where in the digestive system is starch digested?

7 Where are proteins digested?

8 Where are fats digested?

Chemicals are added to your food as it goes through your gut. Your gut is all the organs labelled here in red.

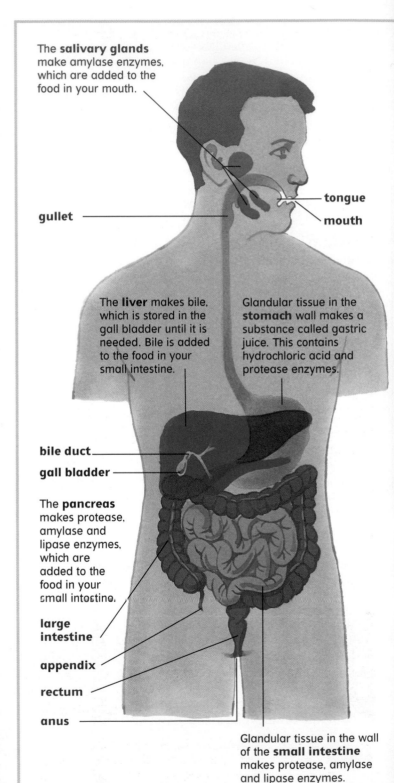

The **salivary glands** make amylase enzymes, which are added to the food in your mouth.

tongue

mouth

gullet

The **liver** makes bile, which is stored in the gall bladder until it is needed. Bile is added to the food in your small intestine.

Glandular tissue in the **stomach** wall makes a substance called gastric juice. This contains hydrochloric acid and protease enzymes.

bile duct

gall bladder

The **pancreas** makes protease, amylase and lipase enzymes, which are added to the food in your small intestine.

large intestine

appendix

rectum

anus

Glandular tissue in the wall of the **small intestine** makes protease, amylase and lipase enzymes.

The stomach produces **hydrochloric acid** because the protease enzymes in the stomach work best in acid conditions. The acid also kills most of the micro-organisms (or microbes) that are in your food.

 9 How does the acid in your stomach protect you?

Bile is added to your food when it gets to the small intestine. The bile **neutralises** (cancels out) the stomach acid and makes your food alkaline. This is because the enzymes that are added to your food in the small intestine work best in **alkaline** conditions.

Bile also helps lipase enzymes to digest fat. It splits large drops of fat into smaller droplets. This is called **emulsification**. Many little droplets have a much larger surface area than one big drop. This gives the lipase enzymes more chance to break down the fats.

 10 Describe two ways that bile helps enzymes to digest fats.

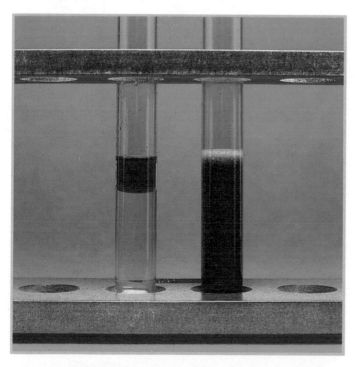

 C *If you shake up a mixture of oil and water, it does not stay mixed for very long. If you put some bile in as well, the oil breaks up into tiny droplets and stays mixed up with the water. This is called an **emulsion**.*

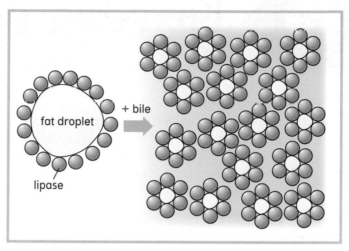

B *When a large fat droplet is split into little droplets, there is more surface area for lipase enzymes to get to the fat and digest it.*

Summary

Enzymes are added to your food as it goes through your gut. Amylase enzymes digest _____, and are made in the salivary _____, the pancreas and the _____ intestine. Protease enzymes digest _____, and are produced by the _____, the pancreas and the small _____. _____ enzymes digest fats. They are produced by the _____ and the small intestine. The stomach also produces _____ acid, which kills _____ and helps enzymes in the stomach to work. Bile is added in the small intestine, which neutralises the _____ and emulsifies fats (_____ them up into small droplets).

---

acid   breaks   glands   hydrochloric
intestine   lipase   micro-organisms
pancreas   proteins   small   starch
stomach

11 You eat a chicken sandwich and drink a glass of milk. Describe the journey of this meal through your gut, explaining which enzymes or other substances are added at each stage, and where the different nutrients are digested.

Enzymes and digestion

How do enzymes help digestion?

Only small, soluble molecules can pass through the walls of your small intestine and get into your blood. Proteins, starches and fats do not dissolve. They have to be split into smaller molecules which do dissolve. Vitamins and minerals do dissolve, and so they do not need to be digested.

 1 a) What is digestion?
b) Name two nutrients that do not need to be digested.

In your body, chemicals called **enzymes** help to speed up or **catalyse** digestion. Enzymes are found in **digestive juices**, which are made in glandular tissue in different parts of the digestive system.

 2 What is glandular tissue? (Hint: see page 10.)

3 a) Which substances help digestion?
b) How do they do this?

4 What are digestive juices?

There are many different kinds of enzymes. Each different kind of enzyme catalyzes the digestion of a different kind of nutrient. The name of an enzyme tells you the nutrient it digests.

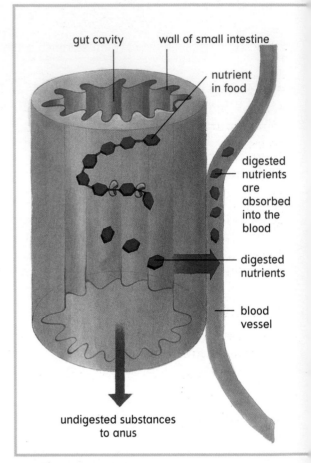

gut cavity   wall of small intestine
nutrient in food
digested nutrients are absorbed into the blood
digested nutrients
blood vessel
undigested substances to anus

A

The names of enzymes always end in -ase. **B**

Nutrient		Enzyme	Product	
	starch	Amylases found in digestive juices from the salivary glands, the pancreas and the small intestine.	glucose (a sugar)	
	proteins	Proteases found in digestive juices from the stomach and the small intestine.	amino acids	
	fats	Lipases found in digestive juices from the pancreas and the small intestine.	fatty acids and glycerol B	

5 Look at table B

   a) What is the name of an enzyme that works on starch?

   b) What are the products of fat digestion?

   c) What are proteins digested into?

Enzymes work best when a mixture is warm. The best temperature is about 37 °C. At cooler temperatures enzymes work more slowly. At high temperatures the enzymes start to fall apart and so cannot work properly.

Each enzyme works best at a certain pH. For instance, the enzymes in your stomach work best at pH 1 (strong acid). If the pH changes an enzyme works more slowly.

6 At which temperature do enzymes work best?

All human enzymes:

- speed up (catalyse) reactions but are not changed themselves
- work on one type of substance only
- work best at 37 °C
- are affected by changes in pH.

7 Table D shows the pH inside different parts of the digestive system. Pepsin is an enzyme that works best in acid conditions. Amylase is an enzyme that works best in neutral conditions.

Part of gut	pH
mouth	7 (neutral)
stomach	1 (strong acid)
small intestine	8 (alkaline)

D

   a) Amylase is produced in the mouth. What pH does it work best at?

   b) Where might pepsin be produced?

   c) Do you think the amylase in the mouth continues to work in the stomach when you swallow your food? Explain your answer.

P Chew a piece of bread for about 3 minutes to mix it with saliva in your mouth.

- What happens to the taste of the bread?
- Can you predict what is happening to the starch in the bread?
- How would you find out if saliva contains an enzyme which digests starch?

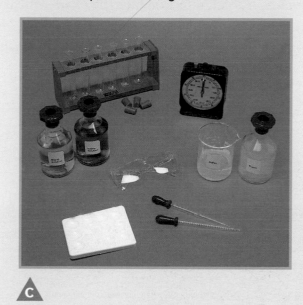

C

Summary

Nutrients often need to be changed into _____ substances so they can get into the blood. This is helped by special chemicals called _____ which _____ your food. Enzymes are found in digestive _____. The enzymes that work on proteins are _____. Amylases digest carbohydrates, like the large molecules of _____. Lipases are enzymes that digest _____. Enzymes work best at _____ °C and each enzyme works best at a particular _____ .

37	digest	enzymes	fats	juices
pH	proteases	soluble	starch	

Diffusion

How do substances move from one place to another?

Your blood carries lots of different substances around your body. When nutrients are **absorbed**, they go through the wall of the small intestine to get into your blood. Dissolved nutrients go from your small intestine into the blood, which goes to your liver.

Water and dissolved substances enter or leave cells by passing through the cell membrane. The membrane will allow some things through but not others.

 1 What enters your blood in your small intestines?

Look at picture B. There is a lot of the purple chemical in the middle of the tank.

Cell membrane.

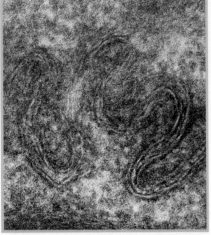

There is a **high concentration** of the purple chemical here

There is a **low concentration** of the purple chemical here

B *The purple chemical has just been put into the tank.*

C *After a few hours, the purple chemical has spread out to fill the tank.*

The particles (or molecules) which make up all substances are moving all the time. Look at picture C. After a few hours, the particles of the water and the purple chemical have moved around and spread out. This spreading movement is called **diffusion**. We say the particles have moved along a **concentration gradient**.

When there is a big difference in concentration, diffusion happens quickly.

 2 What is diffusion?

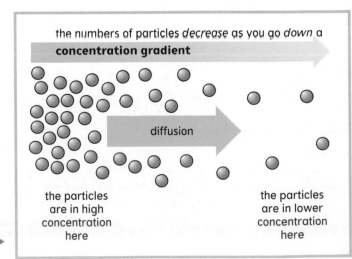

the numbers of particles *decrease* as you go *down* a **concentration gradient**

diffusion

the particles are in high concentration here

the particles are in lower concentration here

D

Substances move into and out of cells in your body by diffusion. There is a high concentration of dissolved nutrients inside your small intestine, and a low concentration in the blood. Nutrients diffuse through the wall of the intestine into the blood.

 3 How do nutrients from food get into your blood?

Nutrients diffuse faster if there is a large surface area for them to diffuse through. The surface area of something can be made larger if it is folded.

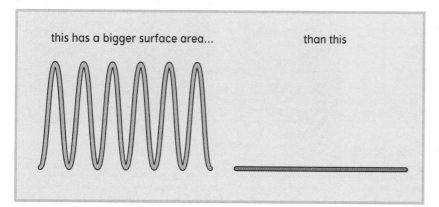

this has a bigger surface area...   than this

F

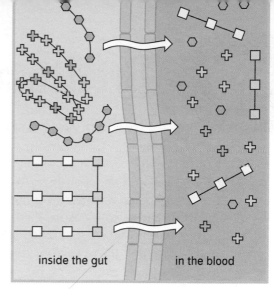

inside the gut   in the blood

 Nutrients diffuse into the blood.

The inside lining of your small intestine has about 5 million tiny finger like projections on it, so the surface is very large. Each projection is called a **villus** and is about 1 mm long. The plural of villus is **villi**.

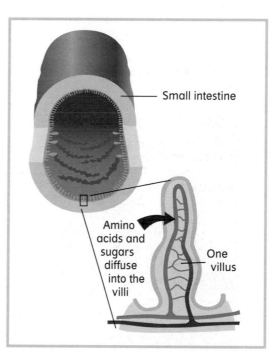

Small intestine

Amino acids and sugars diffuse into the villi

One villus

G

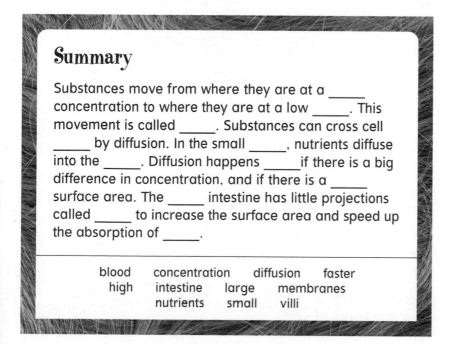

Summary

Substances move from where they are at a _____ concentration to where they are at a low _____. This movement is called _____. Substances can cross cell _____ by diffusion. In the small _____, nutrients diffuse into the _____. Diffusion happens _____ if there is a big difference in concentration, and if there is a _____ surface area. The _____ intestine has little projections called _____ to increase the surface area and speed up the absorption of _____.

blood	concentration	diffusion	faster
high	intestine	large	membranes
	nutrients	small	villi

 4 **a)** What is a villus?
   b) How do villi help nutrients to diffuse into the blood?

5 When you breathe, oxygen diffuses from the air in your lungs into your blood. Where is the concentration of oxygen highest?

6 Diffusion happens faster at higher temperatures. Explain why this happens.

21

Breathing

How do we breathe?

We need to breathe to get oxygen from the air, and to get rid of waste carbon dioxide. When you **breathe**, muscles move to make your lungs bigger and then smaller. This makes air flow in and out of your lungs.

You can survive for 3 weeks without food and 3 days without water but only 3 minutes without air. So in places where there is no air, humans take it with them!

A

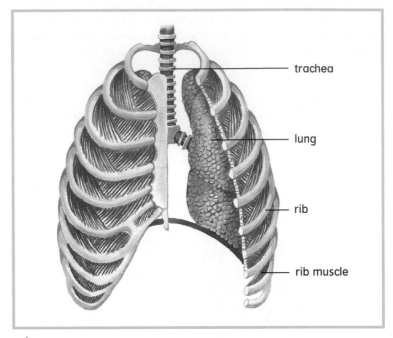

B *Your lungs are protected by your ribs and muscles*

C

trachea

lung

rib

rib muscle

Your lungs are in your **thorax** (chest). The **diaphragm** is a sheet of muscle that separates your thorax from your **abdomen** (the lower part of your body). Your lungs are protected by your ribs. There are muscles between your ribs that can move them.

1 What is the thorax?
2 What is the diaphragm?

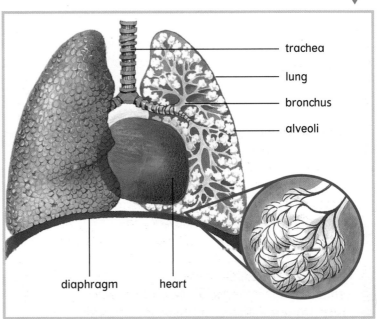

trachea

lung

bronchus

alveoli

diaphragm   heart

When you breathe in, air goes from your nose and mouth, into your windpipe or **trachea**. The trachea divides into two **bronchi**. Each bronchus divides into smaller tubes (**bronchioles**) in your lungs.

At the tips of the smallest branches are tiny pockets (**alveoli**). These alveoli have thin walls and are surrounded by a network of very tiny blood vessels, called **capillaries**, which have blood in them.

3 Name three different tubes that air travels through on its way to the lungs.

When you breathe, the diaphragm and the muscles between your ribs can make your thorax bigger or smaller.

When you breathe in, the movements of the ribs and diaphragm make your thorax bigger. The ribs move out and the diaphragm moves down. Air is sucked in.

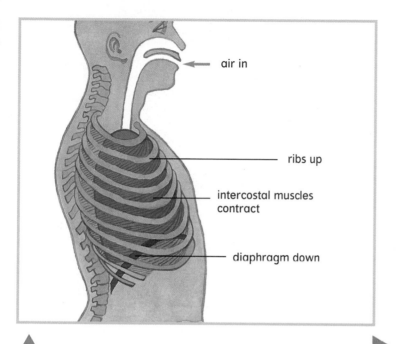

air in

ribs up

intercostal muscles contract

diaphragm down

D

Summary

Your lungs are in your _____, and are protected by your ribs. There are _____ between your ribs which can move them in and out. The _____ is a sheet of _____ at the bottom of the thorax. The rib muscles and the _____ can _____ to make the thorax _____ or smaller. These movements get _____ into and out of the lungs. This movement of air is called _____. Air goes down the _____ into the _____ and then into the bronchioles. The breathing tubes end at alveoli, which are surrounded by _____.

air   bigger   bronchi   capillaries
diaphragm   move   muscle   muscles
thorax   trachea   ventilation

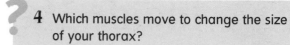

4 Which muscles move to change the size of your thorax?

When you breathe out, your lungs are made smaller. The air inside them flows out of your lungs.

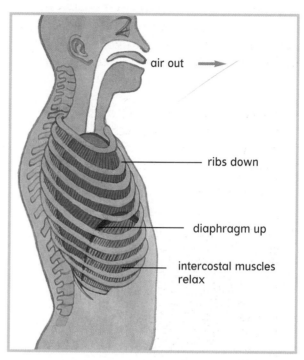

air out

ribs down

diaphragm up

intercostal muscles relax

E

The movement of air in and out of your lungs is called **ventilation**.

You cannot breathe and swallow at the same time. When you swallow, a small flap moves over the opening to your trachea. This flap is your epiglottis. It makes sure food does not go down the wrong way and get into your breathing tubes. If some food lands on the flap, you cough and splutter to jerk the food away.

5 What is ventilation?

6 Air travels from your nose to the alveoli. List the parts it goes through, in order.

7 Explain how you ventilate your lungs.

Inside the lungs

What happens to the air that you breathe in?

Air is moved in and out of the lungs when you breathe. This is **ventilation**. Oxygen from the air needs to get into your blood so that it can be taken to the other cells in your body. Waste carbon dioxide needs to be removed from the blood so you can breathe it out.

If your lungs are to work properly, they must be kept clean. Air enters and leaves your body through your nose and mouth. When you breathe in through your nose, the air passes over the warm, moist lining in your nostrils. The lining is covered in tiny hairs which work like a filter and trap any dust particles that are in the air.

 1 How does your nose help to clean the air you breathe in?

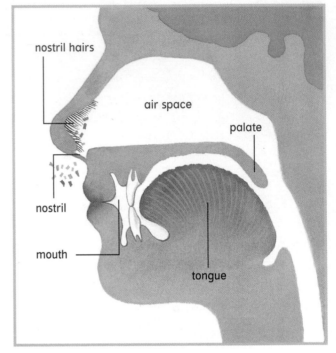

nostril hairs

air space

palate

nostril

mouth

tongue

A

Air moves down the trachea and bronchi into the lungs. There are gland cells in the lining of these tubes. The gland cells make a sticky liquid called **mucus** which traps dust. Other cells have tiny, moving hairs called **cilia**, which move the mucus towards your throat. You swallow this mucus. The mucus you blow out of your nose comes from your nose not your trachea.

 2 Which liquid traps dust?

3 How is the dust moved out of the breathing tubes?

When the air is in the alveoli, oxygen has to **diffuse** into the blood. The walls of the alveoli separate air from the blood in capillaries. The walls of the alveoli and the walls of the capillaries are only one cell thick, so substances like oxygen have only a short distance to diffuse from one place to another.

 4 Why do alveoli have thin walls?

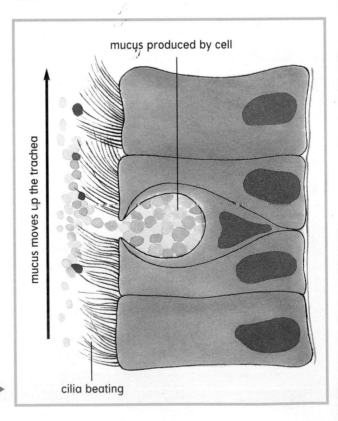

mucus produced by cell

mucus moves up the trachea

B cilia beating

Oxygen is at a higher concentration in the alveoli than in the blood, so it diffuses into the blood. The blood carries the oxygen away, so that the blood around the alveoli always has a low concentration of oxygen.

The blood brought to the lungs from the rest of the body contains waste carbon dioxide, which is at a higher concentration in the blood than in the air in the alveoli. The carbon dioxide diffuses from the blood into the alveoli.

The shape of the alveoli gives the lungs a very large surface area. This helps oxygen and carbon dioxide to diffuse through the alveoli walls quickly.

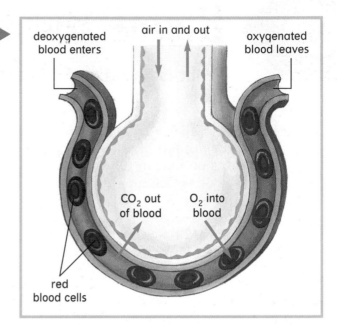

C

deoxygenated blood enters — air in and out — oxygenated blood leaves

CO_2 out of blood

O_2 into blood

red blood cells

5 How does the shape of the alveoli help diffusion?

P What would you predict about the temperature and amount of carbon dioxide in the air you breathe in compared with the air you breathed out?

- Will the air stay at the same temperature?
- Will there be the same amount of carbon dioxide?
- How would you find out?

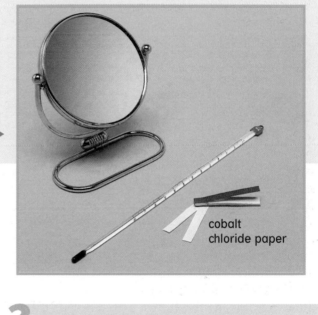

D

cobalt chloride paper

Summary

The air that you _____ in passes through your nose, where _____ trap _____ particles. _____ in your trachea also traps dirt, and tiny hairs called _____ move the mucus and dirt upwards. When the air reaches the alveoli, _____ diffuses into the _____ in the capillaries. Carbon dioxide _____ out of the blood and into the air in the alveoli. The _____ are adapted to their function by having very _____ walls (only one cell thick) and a very _____ surface area.

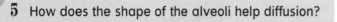

blood	breathe	cilia	diffuses
dust	hairs	large	lungs
mucus	oxygen	thin

6 a) What is the function of the lungs?
b) Describe how they are adapted to this function.

7 It is possible to measure the amount of different substances in the air you breathe in and the air you would breathe out. How would you expect the amounts of the following substances to be different in the two samples? Explain your answers.

a) Oxygen.
b) Carbon dioxide.

Blood

What is blood and what does it do?

Blood looks like a thick, red liquid but it is really a yellow liquid (**plasma**) with lots of red coloured cells in it. Blood travels around your body in tubes called **blood vessels.** The smallest blood vessels are called **capillaries**, which have very thin walls. The blood vessels take the plasma and cells round and round your body, collecting and delivering chemicals. There are 4–5 litres of blood in an adult woman and 5–6 litres in an adult man.

Apart from red blood cells, the plasma also carries white blood cells and platelets. These all have different functions.

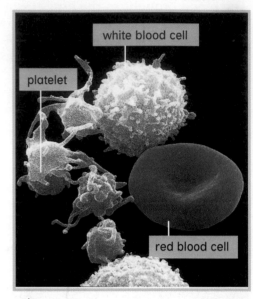

white blood cell

platelet

red blood cell

B

C

	Red blood cell	White blood cell	Platelet
Number in 1 mm³ blood	5 000 000	7 000	250 000
Nucleus	no	yes	no
Life span	120 days	up to 40 years	variable
Job	carries oxygen	protection	blood clotting

A

 1 What is the liquid part of the blood called?

2 What is a capillary?

Red blood cells are very tiny. They are red because they contain a red chemical called haemoglobin. Oxygen attaches itself to haemoglobin when blood flows through the capillaries in the lungs. The red cells carry the oxygen to the cells of your body. Oxygen leaves the haemoglobin and enters the cells that need it.

Your cells use the oxygen to get energy from nutrients such as glucose. Nutrients from your digested food are dissolved in the plasma. As the cells use energy, carbon dioxide and water are produced as waste products. The carbon dioxide leaves your cells and goes into your blood. It is dissolved in the plasma. The carbon dioxide is taken back to the lungs where it passes through the capillary walls into the alveoli and you breathe it out.

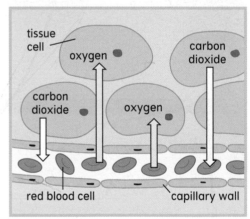

tissue cell

oxygen

carbon dioxide

carbon dioxide

oxygen

oxygen

red blood cell

capillary wall

! Red blood cells do not have a nucleus. A red blood cell survives for about 120 days, so you need to replace them all the time. In the hollow centres of some of your bones there is a substance called bone marrow. This is where your new red cells are made.

 3 What do red blood cells do?

White blood cells are bigger than red blood cells. Your blood has far fewer white blood cells than red blood cells. All white blood cells have a nucleus. These cells can move and squeeze between the cells in the wall of a capillary. The white blood cells help protect you from disease. Some of them can destroy bacteria.

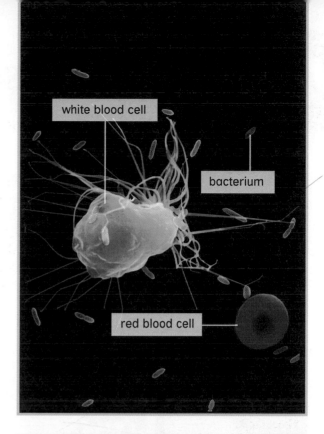

white blood cell

bacterium

red blood cell

4 What do white blood cells do?

5 Give two differences between red blood cells and white blood cells.

Platelets are bits of cells. Platelets do not have a nucleus. The cells they come from are in the bone marrow. The platelets help blood to clot, so that cuts can heal.

6 What do platelets do?

7 Where are platelets formed?

D

A blood clot. E

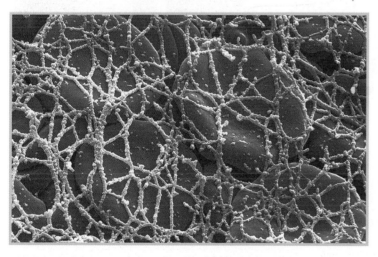

Plasma is mainly made up of water. Chemicals and blood cells are taken around the body in your plasma. Nutrients like glucose and amino acids from the small intestine dissolve in plasma and are supplied to the rest of your body. Carbon dioxide leaves your cells and goes into the blood, where it dissolves in the plasma. The plasma carries it back to your lungs. Your liver breaks down amino acids that your body does not need. This produces waste called **urea**. The urea is dissolved in the plasma and is carried to the kidneys where it is put into the urine and excreted.

8 Write down 3 things that the plasma transports.

Summary

Blood is made of _____ and contains red blood cells, _____ blood cells and _____. Red blood cells carry _____ to the other cells in your body. White blood cells protect the body from _____ and platelets help _____ to heal. Plasma is mainly _____. Plasma _____ many dissolved chemicals around the body.

cuts	disease	oxygen	plasma
platelets	transports	water	white

9 Name a waste chemical that is taken from your liver to your kidneys by the plasma.

10 Capillaries have thin walls. Why do you think this is important?

11 Red blood cells are packed with haemoglobin. Why do you think they have no nucleus?

Blood vessels

Why do you have different kinds of blood vessels?

Your **heart** is an organ that pumps blood through tubes called **blood vessels**. The blood vessels and your heart form the **circulation system**. Blood is pumped from the right side of your heart to the lungs, where it collects oxygen and gets rid of carbon dioxide. This blood then comes back to the heart, where it is pumped to the rest of the body.

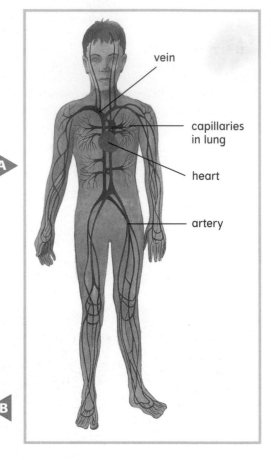

A

vein

capillaries in lung

heart

artery

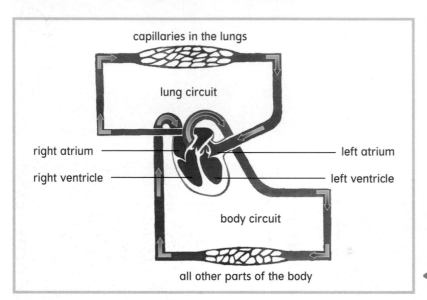

capillaries in the lungs

lung circuit

right atrium

left atrium

right ventricle

left ventricle

body circuit

all other parts of the body

B

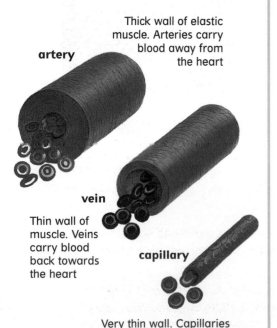

artery

Thick wall of elastic muscle. Arteries carry blood away from the heart

vein

Thin wall of muscle. Veins carry blood back towards the heart

capillary

Very thin wall. Capillaries allow substances to pass in and out of the bloodstream

There are three different types of blood vessels: **arteries**, **veins** and **capillaries**.

Arteries are tubes that take blood away from your heart. Your main artery (the aorta) carries the blood away from the left side of your heart. It is over 2.5 cm wide and branches many times, forming smaller arteries that go to the organs of your body. Another artery leaves the right side of your heart. It divides into two, and one branch goes to your right lung, the other to the left lung. Blood in these arteries picks up oxygen in the lungs and goes back to the left side of the heart. From there the blood goes around your body carrying oxygen to your different organs.

C

1 What are the three different kinds of blood vessel?

2 Where do arteries carry blood from?

Arteries have thick, powerful elastic walls. There is muscle tissue in the wall too and if this contracts, the channel in the artery becomes narrower. This can help to push blood throught the arteries. Blood cells and plasma cannot get through the wall of an artery.

Veins carry blood back to the heart. Veins have thinner walls than arteries, and the walls can stretch easily. Nothing can get through the wall of a vein.

Gravity helps blood get back to your heart from your head. However the blood from your legs has to go upwards against the force of gravity. To stop blood flowing the wrong way, veins often have **valves** in them.

Capillaries are the smallest blood vessels, about the thickness of a hair. In an organ, an artery branches into a network of capillaries. The capillaries pass close to all the cells in your organs. The walls of capillaries are so thin that chemicals in the blood can get through them and reach the cells. The capillaries link up to form a vein which takes blood back to the heart.

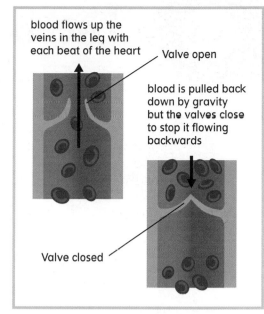

blood flows up the veins in the leg with each beat of the heart

Valve open

blood is pulled back down by gravity but the valves close to stop it flowing backwards

Valve closed

 Vein valves.

A model showing capillaries in a kidney.

3 What is the difference between the wall of an artery and the wall of a vein?

4 Why can't substances get through the walls of arteries and veins?

5 Where do you find capillary networks?

P If you had a piece of artery and a piece of vein, how would you find out which was most elastic (stretchy)?

F

Summary

Arteries take blood _____ from the heart. The wall of an artery is _____. Veins have thinner walls than _____. Veins carry blood to the _____. The _____ link arteries and veins. Capillary walls are thin so substances can pass through. In an organ, capillaries form a _____ of capillaries.

arteries   away   capillaries
heart   network   thick

! Your body contains enough capillaries to stretch around the world one and a half times!

6 What links the artery going to the liver and the vein leaving the liver?

7 List in order the organs and types of blood vessel the blood flows through as it goes from your leg to your lungs.

The heart and its beat

What is your heart and why does it beat?

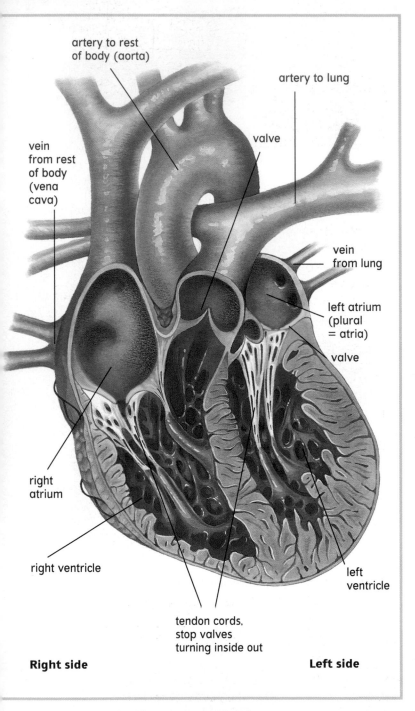

artery to rest of body (aorta)

artery to lung

valve

vein from rest of body (vena cava)

vein from lung

left atrium (plural = atria)

valve

right atrium

right ventricle

left ventricle

tendon cords, stop valves turning inside out

Right side

Left side

A Diagrams of the heart are always drawn as if the person was facing you.

 Your heart beats about 100 000 times each day.

Your heart is held in place inside your chest by strong threads. The heart is made of a special type of muscle which never gets tired. The muscle **contracts** (gets smaller) and pushes blood out of the heart into arteries. When the heart muscle **relaxes** (goes back to its original size) the heart fills with blood from the veins. Each time the muscle contracts, it is called a **heart beat**.

When a heart is cut open, you see a thick wall down the middle. A heart has a right side and a left side; it is two pumps working side by side. Each pump has an upper space called an **atrium**, which receives blood from the veins. Below each atrium is another space called a **ventricle** which pumps blood out of the heart into the arteries.

1 What are the two upper spaces in a heart called?

2 What are the two lower spaces in a heart called?

The two muscular pumps of the heart work together to push blood around your body. The right side pumps blood to your lungs to pick up oxygen and the left side pumps this blood to the rest of your body. Blood travels back to the heart through the veins and the journey round the body begins again. So there are really two circulation systems.

3 Where does the right side of the heart pump blood to?

4 Where does the left side of the heart pump blood to?

An adult heart beats about 70 times every minute. The number of beats each minute is the **heart beat rate**. If you start to move or do any kind of exercise your heart beats faster. This is because during exercise your cells need more oxygen and nutrients. Your heart must work harder to get the blood carrying oxygen and nutrients to your cells quickly.

P How would you find out if your heart beat rate changes when your body is in different positions or doing different activities?

Body position or activity	Pulse rate			mean (average)
	Try 1	Try 2	Try 3	

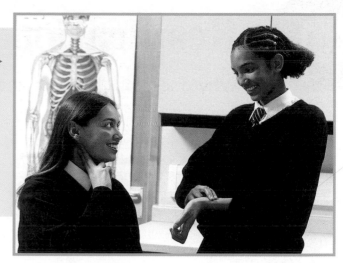
B

Blood is always pushed out of the heart in the correct direction because there are tough **heart valves** between the atria and the ventricles. These valves stop blood flowing back the wrong way.

 5 What stops blood from going the wrong way?

 P How would you find a way of listening to someone's heart? The sounds you hear are the heart valves slamming shut.

C

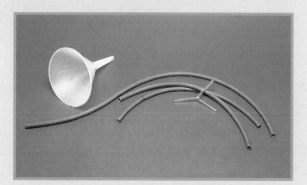

! Your heart does not take nutrients and oxygen from the blood passing through it. It has its own blood supply. The coronary arteries supply the heart muscle with oxygen and food. Waste is taken away in coronary veins. If someone's coronary artery gets blocked, then the heart muscle won't get the oxygen and food it needs. The person will have a heart attack.

D *A partly blocked artery.*

Summary

The heart is in _____ halves. The upper space on each side is called an _____. The spaces beneath are called the _____. Blood flows through the heart in _____ direction. The movement is helped by _____ which stop the blood going in the wrong direction. The heart _____ about 70 times each minute. This is the heart beat _____ .

atrium   beats   one   rate   two   valves   ventricles

? **6** What happens to the heart beat rate when a person exercises?

7 Explain what would happen if a coronary artery was blocked.

8 The left half of the heart has a thicker wall than the right half. Suggest a reason for this.

Respiration

Why does your body need oxygen?

Cells need energy to stay alive. Cells get energy by **respiration**.

The energy from respiration is used

- to make your muscles contract to allow you to move
- to keep you warm
- to help you grow by building up larger molecules from smaller ones.

1 What does your body need energy for?

Aerobic respiration is a chemical reaction that goes on all the time in every cell of your body. **Glucose**, a sugar from digested food, combines with oxygen to release energy. It is called aerobic respiration because it needs oxygen from the *air*. The glucose and oxygen are used up inside the cells and two chemical products are made, carbon dioxide and water. The carbon dioxide is carried away from your cells to your lungs and you breathe it out. Some of the water may be used by the cells.

This is the **word equation** for respiration. Energy is put in brackets because it is not a substance.

glucose + oxygen ⟶ carbon dioxide + water (+ energy)

2 Where does respiration take place?

3 What substances are needed for respiration?

 You use up energy even when you are asleep.

Your muscle cells need glucose and oxygen. When you exercise you need more energy, so your cells need more oxygen and glucose. You breathe faster and deeper to get more air into your lungs so that more oxygen can get into your blood. Your heart beats faster to pump more blood around your body, so that more oxygen and glucose can be carried to your muscles.

P When you exercise, what will happen to the number of breaths you take each minute and the size of those breaths? How would you measure the size of your breaths?

Anaerobic respiration happens when your muscle cells need more oxygen than they can get from your blood. Anaerobic respiration does not need oxygen. A waste product called **lactic acid** is produced.

The word equation is

glucose ⟶ lactic acid (+ energy)

If there is a lot of lactic acid, then your muscles stop working properly and start to ache. After exercise, you breathe deeply for a few minutes. This gets extra oxygen into your blood. The extra oxygen is used to remove the lactic acid and the amount of oxygen needed is called the **oxygen debt**.

? 4 What is anaerobic respiration?

5 When is lactic acid produced?

6 What is the oxygen debt?

This sprinter used anaerobic respiration during her race. She has an oxygen debt.

B

Summary

The normal process which releases energy in cells is called _____ respiration. The waste products are carbon dioxide and _____ . The word equation is

glucose + _____ ⟶ _____ _____ + water (+ _____)

During exercise your heart beat rate and breathing rate increase. Your breaths become _____ and _____ If you don't get enough oxygen, the cells use _____ respiration. This produces _____ _____ . This is removed when extra _____ becomes available. Lactic acid makes muscles _____

> ache   aerobic   anaerobic   carbon dioxide
> deeper   energy   faster   lactic acid   oxygen   water

? 7 What does lactic acid do to your muscles?

8 Why do you pant or take very deep breaths after exercise?

9 What is the difference between ventilation and respiration?

10 Explain why a 100 m runner has a bigger oxygen debt than a marathon runner. (Hint: 100 m runners sprint hard for a short distance, whereas marathon runners run more slowly.)

Microbes and disease

What are the causes of disease?

When a person has a disease their body is not working properly. There are signs that a doctor can see or find out using instruments or by doing a test. These signs or **symptoms** include things like a high temperature, a skin rash or the wrong sort of chemicals in the urine.

Some diseases are caused by not eating properly. Some diseases are linked to growing old, and others may be passed on by our parents. A large number of diseases are caused by **microbes** (short for **micro-organisms**). Microbes are tiny organisms that can only be seen with the help of powerful microscopes.

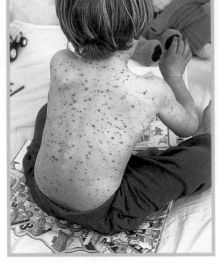

A   *Chickenpox*

? 1   What is a symptom?

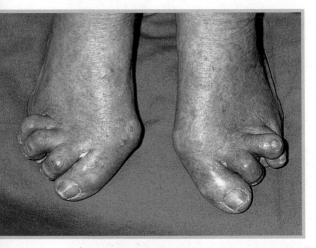

B   *Arthritis*   *Aids*   **C**

Bacteria are single cells that can only be seen through a microscope. The cells have cytoplasm and cell membranes which are surrounded by a white cell wall. These cell walls are flexible – not like the cell walls in plant cells.

Bacteria have a coiled up strand of **DNA** (a **chromosome**) which contains **genes**. The genes are not in a nucleus. Bacteria reproduce when a single cell splits into two new ones.

We use bacteria in many ways; for example, we mix them with milk to make yoghurt. A few bacteria are harmful and cause disease either by destroying our cells or making poisons called **toxins**.

? 2   How do bacteria reproduce?

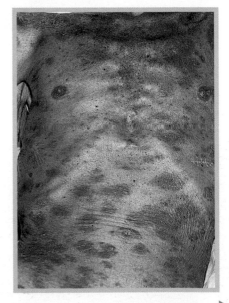

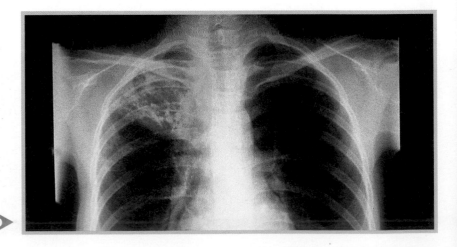

Tuberculosis   **D**

Disease caused by bacterium	Symptom
Tuberculosis	coughing up blood.
Impetigo	red patches on skin.
Cholera	fever, diarrhea and thirst.

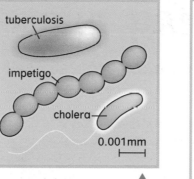

tuberculosis

impetigo

cholera

0.001mm

 E

Viruses are microbes that are much smaller than bacteria. Viruses can only live and reproduce inside another living cell and so they damage the cells they live in. A virus has a layer of protein on the outside — a **protein coat**. Inside this is a short strand (often made of DNA) containing genes.

 G

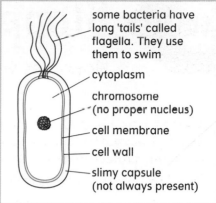

some bacteria have long 'tails' called flagella. They use them to swim

cytoplasm

chromosome (no proper nucleus)

cell membrane

cell wall

slimy capsule (not always present)

F *General structure of bacteria.*

General structure of a virus. H

Disease caused by virus	Symptom
Mumps	Salivary glands swell.
Chickenpox	Small spots forming a rash on the skin.
AIDS	May stop the immune system from working, so the person gets other diseases.

Mumps

Chickenpox

AIDS

0.00001mm

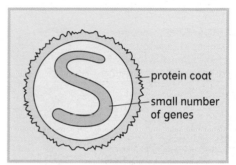

protein coat

small number of genes

! In the right conditions, a bacterium can divide once every 20 minutes. So in one day, a simple bacterium could produce 1000 million, million, million offspring! There are more bacteria in your body than there are people on earth, yet they could all be packed into a soup tin.

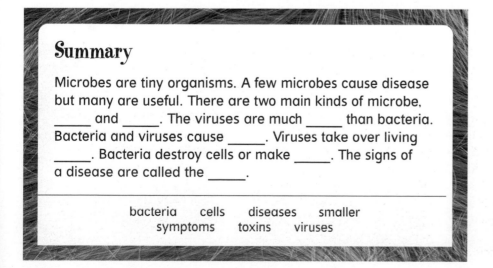

Summary

Microbes are tiny organisms. A few microbes cause disease but many are useful. There are two main kinds of microbe, _____ and _____. The viruses are much _____ than bacteria. Bacteria and viruses cause _____. Viruses take over living _____. Bacteria destroy cells or make _____. The signs of a disease are called the _____.

bacteria   cells   diseases   smaller
symptoms   toxins   viruses

?

3 Where do viruses reproduce?

4 Which is bigger, a virus or a bacterium?

5 Look at table E. List the symptoms of
   a) tuberculosis
   b) impetigo.

6 Write down the names of
   a) two diseases caused by bacteria
   b) two diseases caused by viruses.

7 How are bacteria useful to us?

8 What is the job of the flagellum of a bacterium?

9 Give 3 differences between a bacterium and a virus.

Stopping microbes

How does your body protect itself from microbes?

Your body is a battleground. Microbes settle on it and try to break in, but your body usually fights back and protects you from disease.

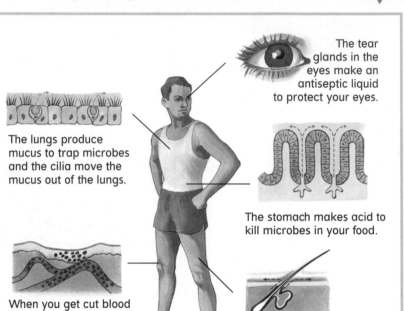

A

The tear glands in the eyes make an antiseptic liquid to protect your eyes.

The lungs produce mucus to trap microbes and the cilia move the mucus out of the lungs.

The stomach makes acid to kill microbes in your food.

When you get cut blood clots form scabs to stop microbes entering the body.

Glands in your skin make antiseptic oils. The skin is a protective layer of dead cells.

Your skin protects you. It keeps out microbes. But cuts and grazes can open the way for microbes to get in. Some parts of your body, like your eyes, are not protected by skin. Tear glands produce an **antiseptic** liquid that kills microbes. Sweat and saliva also contain these chemicals.

Your throat is a direct route into your body and you swallow microbes with your food. Cells in your stomach make hydrochloric acid and this kills most microbes.

Your nose is a direct route into your body. The lining of the breathing tubes are lined with sticky **mucus** which traps most microbes. The cilia inside the breathing tubes sweep the mucus and trapped microbes up the trachea so that you can swallow it. The acid in your stomach kills the microbes.

? 1 How are your eyes protected from infection?

2 What happens to microbes you swallow with your food?

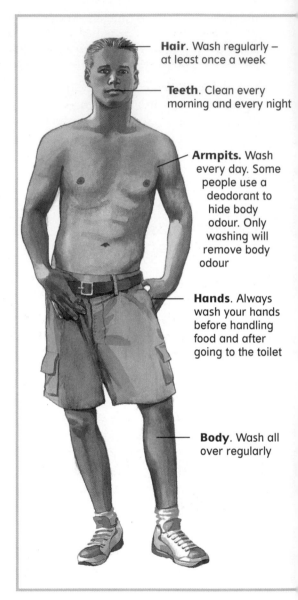

Hair. Wash regularly – at least once a week

Teeth. Clean every morning and every night

Armpits. Wash every day. Some people use a deodorant to hide body odour. Only washing will remove body odour

Hands. Always wash your hands before handling food and after going to the toilet

Body. Wash all over regularly

B *Simple hygiene rules.*

You can cut down the chances of becoming infected by taking care of your body as shown in diagram B.

? 3 Why is it important to wash your hands after going to the toilet?

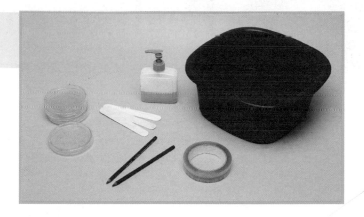

The food we prepare for ourselves can also be food for microbes. Some of these microbes are harmless. Some can cause a disease, like **food poisoning**, if they enter our bodies. One common form of food poisoning is caused by a bacterium called *Salmonella*. People often get *Salmonella* food poisoning from chicken or eggs that are not properly cooked.

! Botulism is the most dangerous type of food poisoning. It is caused by bacteria called *Clostridium botulinum*. This microbe makes a toxin that is the most poisonous substance known. Two or three spoonfuls would be enough to kill 100 million people!

If someone has a disease caused by a virus or a bacterium, they are **infected**. They can pass the disease on to other people. Different diseases spread in different ways. Some are spread by touching an infected person, or by breathing in microbes.

Microbes from infected people or animals can get into drinking water. If the water is not cleaned properly, people who drink it can catch the disease.

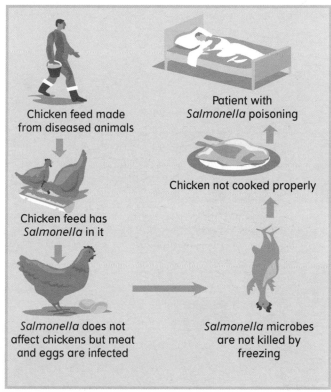

Chicken feed made from diseased animals

Chicken feed has *Salmonella* in it

Salmonella does not affect chickens but meat and eggs are infected

Patient with *Salmonella* poisoning

Chicken not cooked properly

Salmonella microbes are not killed by freezing

D *How food poisoning spreads.*

Summary

The surface of your body is protected by your _____. Chemicals in _____ protect your eyes. Many of the microbes in the air are trapped by _____ in your breathing tubes. Hydrochloric _____ in your _____ kills microbes you may swallow. Food may contain _____ that cause food _____. Diseases can be passed on by touching an _____ person, or by breathing in microbes. Diseases can also be passed on in _____ water.

acid   bacteria   drinking   infected
mucus   poisoning   skin   stomach   tears

?

4 Write down 3 hygiene rules for both men and women.

5 Write down two ways that you can catch a disease from an infected person.

6 Why should you not drink water out of a river?

7 Explain why it is important to cook chicken properly.

Healing cuts

How does your body heal cuts?

When you bleed, the flow of blood helps to wash away any microbes that get into the cut. A wound is sealed by a **blood clot**, which also stops more microbes getting in and blood getting out. The clot of blood dries, forming a **scab**. New skin grows under the scab to seal the cut permanently.

 1 What is a scab?

If someone suffers from a disease called haemophilia, their blood does not clot properly. The disease can be passed on by parents to their children. Alexis, the son of Russian Tsar Nicholas II, was a haemophiliac. If Alexis was cut, his wounds bled a lot and for a long time.

B

Alexis

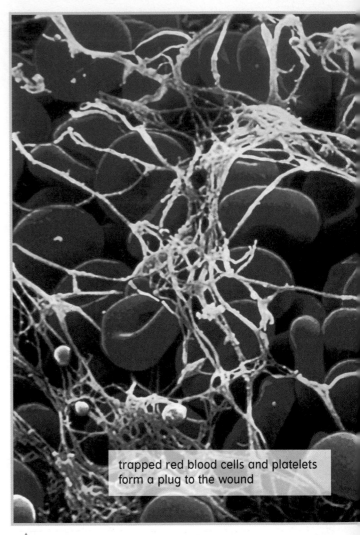

trapped red blood cells and platelets form a plug to the wound

 A blood clot.

If microbes get through the body's defences, the microbes will be in a warm place. In warm conditions the microbes multiply quickly and attack the cells or put toxins (poisons) into the blood. Your body can identify these dangerous microbes and destroy them without damaging your own cells. This is the job of the **immune system**. The white blood cells are the most important part of this system.

 2 What is the job of the immune system?

Some white blood cells destroy microbes by surrounding them. A white cell is said to **ingest** the microbe. At a cut or a 'spot', there may be many white cells fighting the infection. These cells and the dead microbes form a yellow **pus**. Other white blood cells make **antitoxins**. These are chemicals which cancel out the **toxins** (poisons) made by microbes.

3 How do white blood cells protect you?

4 What is pus?

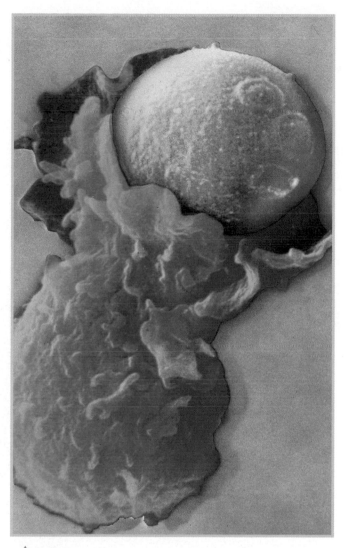

C   *White blood cells surrounding a yeast cell.*

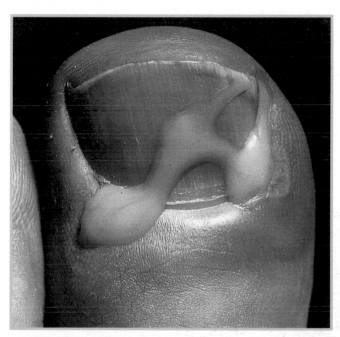

D   *Pus from an infected toe.*

When someone is badly injured they can lose a lot of blood. Blood is made inside the body to replace what is lost, but this takes time. Blood can be taken from a healthy person (a donor) and given to an injured one in a blood transfusion. Doctors must check that the blood of the donor matches the blood of the person who needs it.

5 What are antitoxins?

6 Explain why there may be pus in a spot.

7 Explain why someone may need to have a blood transfusion.

Summary

When a wound bleeds, _____ are washed away. The wound is sealed by a blood _____ which dries to form a _____ . If microbes get into a cut, they _____ quickly and may make _____ . The _____ system helps destroy microbes and toxins.

| clot | immune | microbes |
| multiply | scab | toxins |

Immunity

Why do we catch some diseases and not others?

Lots of children catch diseases like measles, mumps, chickenpox and whooping cough. Adults rarely get them. Other diseases are found in animals but not humans. These differences are due to the way our **immune systems** work.

Some microbes cause diseases in animals. For example, one virus that attacks cattle causes a disease called 'foot and mouth' disease. This virus cannot grow inside human cells. This means we cannot catch this disease. We are naturally protected or **naturally immune**.

Animals with 'foot and mouth' disease have to be killed and burned to stop the disease from spreading.

?1 Why don't humans get 'foot and mouth' disease?

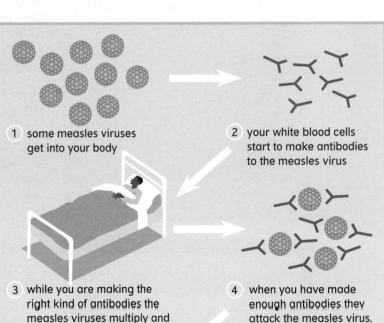

1 some measles viruses get into your body

2 your white blood cells start to make antibodies to the measles virus

3 while you are making the right kind of antibodies the measles viruses multiply and make you ill

4 when you have made enough antibodies they attack the measles virus, sticking them together

5 the measles viruses are made harmless and you get better

6 if the measles virus gets into your body again your white blood cells know what kind of antibody to make right away and they can kill it before you get ill - you are now immune to measles

The first time that measles microbes get into your body, you get measles. Your white blood cells take time to recognise the measles microbes. Once they have recognised the microbes, they produce **antibodies**. Antibodies kill the microbes. If the measles microbe gets into your body again, your white blood cells already have a 'plan of action' and can produce the antibodies very quickly. They do not need to spend time recognising the microbes. This is why you can only get measles once. Once you have had measles, you have become **immune** to the disease. The measles antibody works only against measles, it does not protect you against other microbes.

?2 What is an antibody?

3 What does being immune to measles mean?

4 A measles antibody will not protect you from mumps. Why not?

Doctors can help to protect you from diseases. When you have a 'jab' or injection a small dose of a weak or dead microbe is put in your body. The small dose makes your body produce the antibodies that fight the infection. Later you may be given another jab that causes your body to make even more antibodies. This is the booster jab. You will be protected for a long time against the disease. Injections like this are called **vaccinations**. The weak or dead microbe you are given is called a **vaccine**. As a vaccine makes you immune to a disease, having a vaccination means you are **immunised**.

Summary

It is the job of the _____ system to protect the body from disease. The white blood cells produce _____ which kill the microbes. Once the white blood cells have made an antibody, they can make more antibodies easily. Doctors can give you an injection called a _____ to protect you against some diseases. The injection will contain a weak or _____ form of the _____ . Once you have had the injection your body is _____ against the disease.

antibodies	dead	immune
immunised	microbe	vaccination

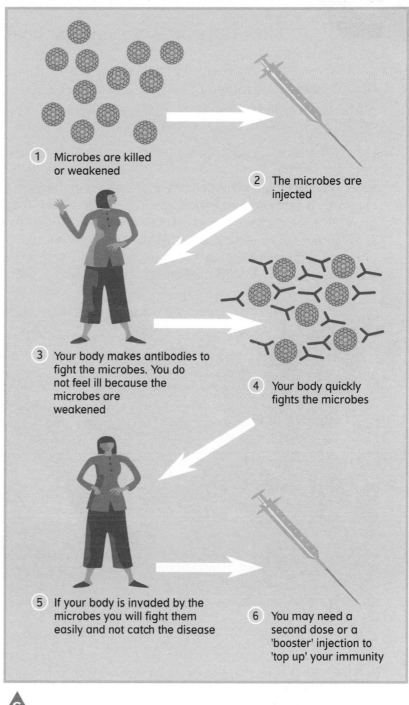

1. Microbes are killed or weakened
2. The microbes are injected
3. Your body makes antibodies to fight the microbes. You do not feel ill because the microbes are weakened
4. Your body quickly fights the microbes
5. If your body is invaded by the microbes you will fight them easily and not catch the disease
6. You may need a second dose or a 'booster' injection to 'top up' your immunity

5 If you have measles once, you are unlikely to catch it again. Explain why.

6 Explain why a second, or booster, jab is needed to protect you against tetanus for a long time.

Health and lifestyle

How can you keep yourself healthy?

Not all diseases are caused by microbes. Some are caused by eating the wrong kinds of food, or by smoking. Your **lifestyle** can affect your health, and exercise is important.

Your heart beats about 70 times each minute. Some fit athletes have hearts that pump as slowly as 40 times a minute. Athletes train and this makes the heart work harder. The heart gets bigger so it pumps more blood at each beat and rests longer between beats. When the athlete is at rest, the trained heart beats more slowly than normal – but it is still working properly. A slow heart rate is one sign of fitness. You can help to keep your heart healthy by getting plenty of exercise.

Athletes have to be very fit. **A**

 1 How can you keep your heart healthy?

P How would you investigate what the effect of exercise is on your heart rate? Is it the same as the other pupils in your class? **C**

B *Playing sports keeps you fit but even walking to school helps you to stay healthy.*

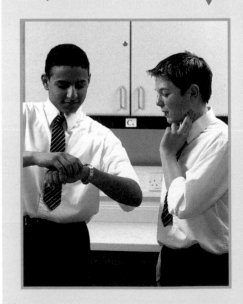

Healthy arteries have a flexible wall and a smooth lining. In some people, the arteries change as they get older. A fatty substance, called cholesterol, starts to stick to the lining. Then the artery walls become thicker and harder. This means the blood supply to an organ will be less. If this organ is the heart, a person may have chest pains when they try to run. If the arteries become blocked, it causes a heart attack and possibly death.

Cholesterol is found in red meat, eggs and milk. You need some of the cholesterol to stay healthy, but too much leads to heart disease. You can keep your arteries healthy by eating the right kind of diet.

 D This meal contains a lot of cholesterol.

This meal still contains fat, but not as much. **E**

You would expect someone with a diet of milk and meat to have a lot of cholesterol in their blood and so run the risk of having a heart attack. The Masai people of Kenya have a diet like this and yet they stay healthy. This is because they eat a plant that contains chemicals called saponins. These have been found in soya beans and chick peas. Saponins stop cholesterol moving across the gut wall into the blood. **F**

2 What is a heart attack?

3 How can you protect yourself against a heart attack?

Sometimes a lump of fat forms a clot inside a blood vessel. If a clot gets stuck in an artery or vein it will stop the flow of blood. A clot in the brain cuts down the blood supply to the brain. This lowers the amount of oxygen reaching the brain cells and can damage the cells or even kill the person. This is called a stroke.

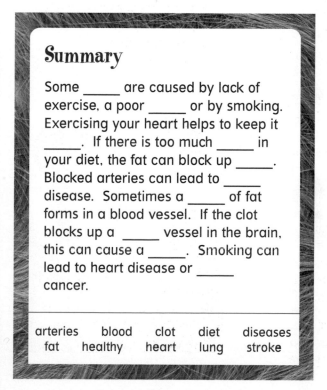

4 What is a stroke?

Smoking is also bad for you. Chemicals in tobacco smoke can cause lung cancer. The chemicals also affect your blood vessels, and can lead to heat attacks or strokes.

5 Why is smoking bad for your health?

You can keep yourself healthy by having a healthy lifestyle.

Summary

Some _____ are caused by lack of exercise, a poor _____ or by smoking. Exercising your heart helps to keep it _____. If there is too much _____ in your diet, the fat can block up _____. Blocked arteries can lead to _____ disease. Sometimes a _____ of fat forms in a blood vessel. If the clot blocks up a _____ vessel in the brain, this can cause a _____. Smoking can lead to heart disease or _____ cancer.

| arteries | blood | clot | diet | diseases |
| fat | healthy | heart | lung | stroke |

6 Design a leaflet to tell people how they can change their lifestyle to keep them healthy. You could also include some information from topic A16 (page 36).

43

1 a) Copy and complete these sentences using words from the box. You may use each word once, more than once, or not at all.

All living things are made of ____. A group of ____ working together is called a ____. Several different tissues may work together in an ____. An ____ ____ carries out a particular ____ and may have several different organs. (7)

cells	cytoplasm	function	nucleus
organ	skin	system	tissue

b) What is the function of glandular tissue? (1)

c) i) Name three organs in the digestive system. (3)

ii) What does the digestive system do? (2)

2 This diagram shows an animal cell.

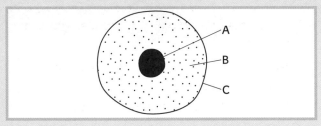

a) i) What is part A?   **ii)** What does it do? (2)

b) i) What is part C?   **ii)** What does it do? (2)

This diagram shows a ciliated epithelial cell.

c) What is the function of this kind of cell? (1)

d) Where in the body would you find this kind of cell? (1)

e) Describe one way in which this cell is adapted to its function. (1)

3 This diagram shows part of the digestive system.

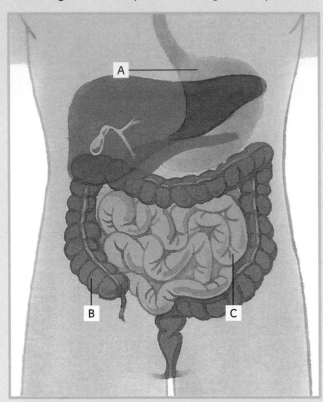

a) Write down the names of parts A, B and C (3)

b) In which of these parts are nutrients absorbed into the body? (1)

c) What does part A do?

4 Various substances are added to your food as it moves through your digestive system. Describe **two** functions of each of these:

a) saliva (2)

b) acid in the stomach (2)

c) bile. (2)

5 When you breathe in, the air goes through all the following parts of your breathing system. Write them out in order, starting with your nose. (3)

bronchi	alveoli	trachea
	nose	bronchiole

6 Copy and complete these sentences using words from the box. You may use each word once, more than once, or not at all.

Your ____ form part of your ____ system, which moves ____ in and out of your body so that ____ can diffuse into your blood. The lungs are in the ____ and are protected by ____.
When you breathe in, ____ between your ribs pull your ribs ____, and your ____ becomes flatter. These changes make your thorax ____, so air flows into your lungs. When you relax these ____, your ____ become ____, and air is forced out. (13)

abdomen	air	bigger	breathing
diaphragm	lungs	muscles	outwards
oxygen	ribs	smaller	stomach	thorax

7 This diagram shows a heart.

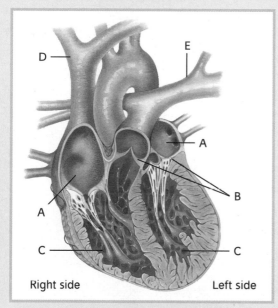

Right side   Left side

a) i) What are the spaces labelled A called?
ii) What are the spaces labelled C called? (2)

b) What is the function of the structures labelled B? (1)

c) i) Is blood vessel D an artery or a vein?
ii) Explain your answer. (2)

d) i) Which side of the heart will contain blood with a lot of oxygen in it?
ii) Explain your answer. (2)

e) Why is the left side of the heart bigger than the right side? (2)

8 **a)** Copy and complete the word equation for aerobic respiration. (3)

____ + oxygen ⟶ ____ ____ + water (+ ____)

b) i) What form of respiration happens in cells when they do not get enough oxygen?
ii) What is the waste product from this process?
iii) What is the oxygen debt? (3)

9 Describe one function of each of these parts of the blood.
a) white blood cells (1)
b) red blood cells (1)
c) plasma (1)
d) platelets. (1)

10 Read the passage and then answer the questions.

You can protect yourself against infections by washing your hands before preparing food, by storing and cooking food properly, and by getting vaccinated against common diseases like measles and tetanus.

You can keep your body healthy by getting plenty of exercise, by eating a healthy diet, and not smoking.

a) Name two kinds of microbe that can cause infections. (2)

b) Why is it important to cook food properly? (2)

c) Describe two ways that your body protects itself against microbes. (4)

d) What is a vaccination, and how does it protect you against infection? Explain as fully as you can. (3)

e) Name the five kinds of nutrient that humans need. (5)

f) Describe one harmful effect that an unhealthy diet can have. (1)

Plant structure

What are the features of flowering plants?

A We can eat the roots of carrot plants.

B We can eat the leaves of spinach plants.

C We can eat the stems of celery plants.

D We can even eat some flowers. This is broccoli.

Plants provide many useful substances for us, including food. We eat many different parts of different plants.

1 Write down four different parts of plants that we can eat.

Not all plants can be eaten. All the parts of the oleander plant are deadly poisonous. People have died by eating food cooked over oleander wood. Bees make poisonous honey if they visit too many oleander flowers. **E**

The main parts of a plant are called **plant organs** and each organ has a particular job to do.

2 Look at picture F.
   a) Write down the names of four plant organs.
   b) Which organ on your list makes food for the plant?

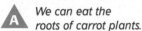
The main organs of a flowering plant. **F**

Many plants have **flowers**. They make seeds from which new plants grow. They also contain the male and female **sex organs**.

The **leaf** is used to make food for the plant. Leaves use **photosynthesis** to do this. Photosynthesis needs light energy to make it work.

The **stem** helps to support the plant and hold the leaves in place. It also contains tubes which carry water and food around the plant.

The **roots** keep the plant in the ground and take in water and mineral salts from the soil. Mineral salts are important for healthy growth.

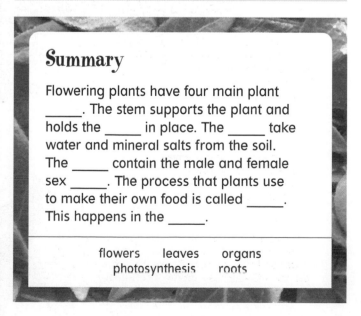

G Some flowers are not obvious. These are grass flowers.

Plants which have flowers are known as **flowering plants**. Flowers come in many different shapes and sizes. They produce **seeds** from which new plants will grow. The seeds of flowering plants are often contained in **fruits**. Any plant organ that is eaten and is not a fruit, is called a **vegetable**.

H Many fruits are useful. Chocolate is made from the seeds found inside the fruit of the cacao tree.

? 3 a) Make a list of 5 vegetables.
 b) For each vegetable on your list, write down what plant organ it is.

? 4 Picture I shows a tomato. Write down whether a tomato is a fruit or a vegetable. Explain your answer.

I

Summary

Flowering plants have four main plant _____. The stem supports the plant and holds the _____ in place. The _____ take water and mineral salts from the soil. The _____ contain the male and female sex _____. The process that plants use to make their own food is called _____. This happens in the _____.

flowers   leaves   organs
photosynthesis   roots

5 What do flowers do?

6 Why do you think plants die if they do not get any light?

7 Draw a diagram of a plant. You can either draw a real plant or make one up. Label the organs.

Plant cells

What are the features of plant cells?

All living things are made out of smaller units called **cells**. Robert Hooke discovered cells in the middle of the 17th century. Using a **microscope** he looked at thin layers of cork and observed small boxes which he called cells.

A monk's cell

1 What piece of equipment did Hooke use to see cells?

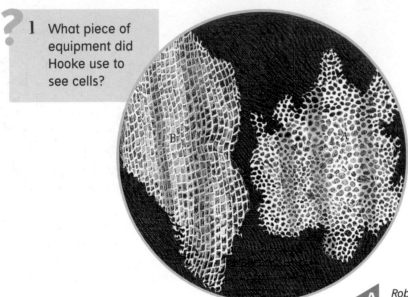

A Robert Hooke's drawing of cork cells from his book Micrographia *published in 1665.*

Hooke used the term 'cells' because he thought they looked a bit like the small rooms (or cells) that were found in monasteries.

B

Plant cells have some features that are different from animal cells and they look more like 'boxes' than animal cells.

2 **a)** Draw a plant cell and label the parts.
b) On your drawing, draw a box around the labels for parts that *only* plant cells have.

C

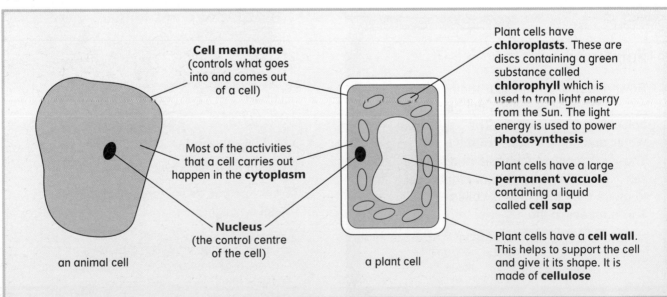

Cell membrane (controls what goes into and comes out of a cell)

Most of the activities that a cell carries out happen in the **cytoplasm**

Nucleus (the control centre of the cell)

an animal cell

a plant cell

Plant cells have **chloroplasts**. These are discs containing a green substance called **chlorophyll** which is used to trap light energy from the Sun. The light energy is used to power **photosynthesis**

Plant cells have a large **permanent vacuole** containing a liquid called **cell sap**

Plant cells have a **cell wall**. This helps to support the cell and give it its shape. It is made of **cellulose**

Picture D shows some cells from an onion. The cells have had a **stain** added to them so that the parts can be seen better.

3 a) Which part, normally found in plant cells, is not found in these onion cells.

b) Write down why you think this part is missing.

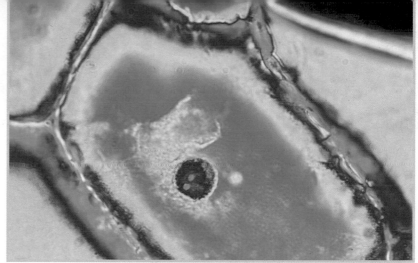

Stained onion cells as seen through a microscope.

P How would you prepare some onion cells so that you could see them? **E**

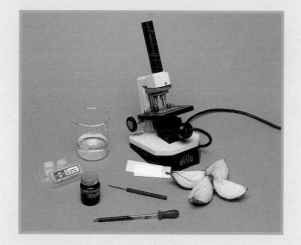

4 a) What substance do chloroplasts contain?
b) What colour do you think this substance is?
c) What does this substance do?

5 a) What are cell walls made out of?
b) From your knowledge of cell walls, name one property that this substance should have.

6 In 1882 Walther Flemming published a book describing cells in detail for the first time. What do you think Walther Flemming used to see cells more clearly?

7 When examining onion cells it is often easier to use red onions rather than the white ones. Why do you think this might be?

Summary

Name of part	The job it does	Is it found in animal cells?	Is it found in plant cells?
	Contains cell sap.		
	Controls the cell.		
	Controls what goes into and out of the cell.		
	Helps support the cell.		
	Most of a cell's activities happen here.		
	Photosynthesis happens here.		

cell membrane   cell wall   chloroplasts   cytoplasm   nucleus   vacuole

Adaptations of plant cells

How are plant cells adapted to do certain jobs?

Car designers need to design different cars for different jobs. The photographs show some newly designed cars. The job (or function) of car A is to be able to go off road and it has features to allow it to do this. We say that the car is **adapted** to its **function**.

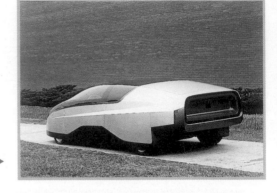

 Car A — an off-roader.

Car B — a sports car. **B**

? **1** Write down one feature of car B that makes it adapted to its function.

Cells also have different features for different functions.

Palisade cells

Palisade cells are found near the top of leaves and so they get a lot of light from the Sun. They contain many chloroplasts to trap as much light energy as possible for photosynthesis.

D

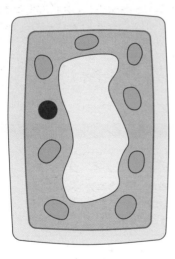

The function of palisade cells is to produce food using photosynthesis. They are adapted to do this by having many chloroplasts

C

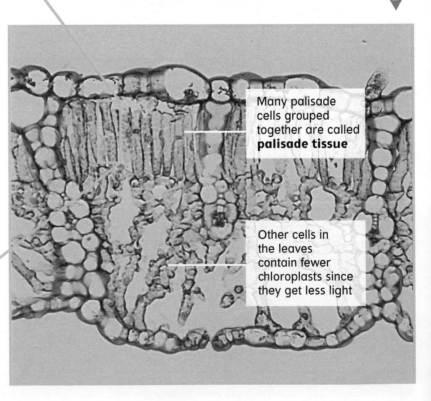

Many palisade cells grouped together are called **palisade tissue**

Other cells in the leaves contain fewer chloroplasts since they get less light

? **2** How are palisade cells adapted to their function?

Root hair cells

The function of **root hair cells** is to take water and mineral salts out of the ground. They are adapted to this function by having 'root hairs'. A root hair helps the cell take in water by increasing its surface area.

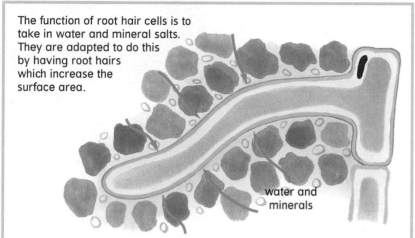

The function of root hair cells is to take in water and mineral salts. They are adapted to do this by having root hairs which increase the surface area.

water and minerals

E

Many root hair cells grouped together are called **root hair tissue**

F

3 Make a drawing of a root hair cell and label its parts.

P How would you find out if increasing the surface area of a sponge helps it to soak up water more quickly?

4 a) Explain what is meant by surface area.
   b) Explain how a root hair cell is adapted to its function.

5 Picture G shows the top and bottom sides of a leaf. Why do you think the bottom side looks less green?

G *Top and bottom sides of a leaf*

6 Using the information on this page, explain what is meant by a tissue.

Summary

Different cells have different features for different jobs. We say that cells are _____ to their _____. _____ cells are important for making food for the plant. They have many _____ to help them do this. _____ _____ cells take _____ and _____ _____ out of the soil. The root hairs increase the _____ _____ of the root.

adapted   chloroplasts   functions
mineral salts   palisade   root hair
surface area   water

Roots and stems

What are the tissues in roots and stems used for?

Many people like eating celery because it is crunchy. Its crunchiness is due to cells in the stem called **xylem cells**.

Xylem cells are dead cells that are adapted to carry water and mineral salts to where they are needed in the plant.

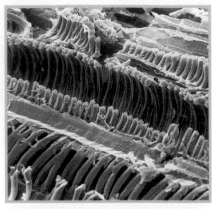

 Xylem tissue as seen through a very powerful microscope. The xylem tubes have been cut in half.

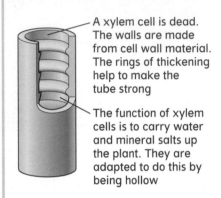

A xylem cell is dead. The walls are made from cell wall material. The rings of thickening help to make the tube strong

The function of xylem cells is to carry water and mineral salts up the plant. They are adapted to do this by being hollow

 A This stem of celery has been left in a blue dye for a few hours. The dye has risen up through the xylem tubes. **B**

? **1** How is the structure of a xylem cell adapted for its function?

The dead xylem cells join together forming long tubes, like thin straws. When cells of the same type are found together, they are known as a **tissue**. Picture C shows **xylem tissue**.

? **2** What do you think a large group of palisade cells in a leaf is called?

3 Draw a diagram to show some root hair tissue.

An **organ** contains different tissues, working together to do an important job. The root and stem are plant organs which both contain xylem tissue. They also contain strands of living cells (**phloem tissue**) that carry food from the leaves to all the other parts of the plant. The food is needed by the growing parts of plants. The food can also be moved into **storage organs** to be stored. Potatoes are storage organs.

The root

Root hair tissue takes water and mineral salts out of the soil. These are moved into xylem tissue. Water moves from the soil into root hair cells by **osmosis**. The cell membrane is **semi-permeable**, which means that water molecules can go through it but most larger molecules cannot.

When a little water contains lots of mineral salts, the solution of mineral salts is said to be **concentrated**. If there is a lot of water, the solution of mineral salts is said to be **dilute**. Water moves by **osmosis** from a dilute solution to a concentrated solution.

Growing parts

Potatoes are storage organs

 D A potato plant.

The solution of mineral salts inside a cell is concentrated. In the soil the solution is dilute. So water moves from the soil into the plant. The cells nearer the xylem have more and more concentrated solutions in them and so water moves from cell to cell and into the xylem.

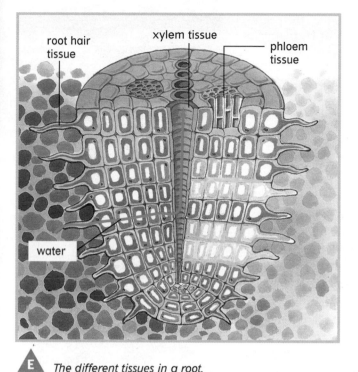

root hair tissue

xylem tissue

phloem tissue

water

E The different tissues in a root.

Summary

A group of cells of the same type and doing the same job is called a _____.
An _____ is a group of different tissues working together to do one important job. _____ cells are adapted to carry water by having a _____ middle. A group of these cells is known as _____ _____ . _____ cells are used to carry food up and down the plant.

Water moves from cell to cell through cell membranes by _____. The water moves from cells where the solution inside the cell is _____ to cells where the solution inside the cell is _____.

concentrated	dilute	hollow
organ	osmosis	phloem	tissue
	xylem	xylem tissue	

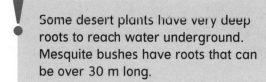

Some desert plants have very deep roots to reach water underground. Mesquite bushes have roots that can be over 30 m long.

4 a) What is the name of the tissue used to carry food down to the roots?
b) Where is this food made?

5 What is osmosis?

The stem

The stem transports substances up and down the plant and helps to support the plant. Xylem tissue is mainly made out of **cellulose** which is very strong. Tall plants, like trees, have stems with a lot of xylem tissue in them.

6 Write down two functions of the stem.

7 Why do you think tall plants have stems containing a lot of xylem?

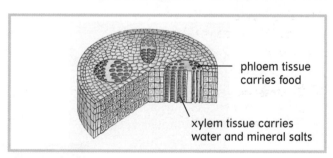

phloem tissue carries food

xylem tissue carries water and mineral salts

F The different tissues in a stem.

8 Write down the names of two plant organs that contain xylem cells.

9 Some plants have green stems. What would you expect to find in the cells of these stems?

10 Food is made in the leaves.
Some of it needs to reach the roots.
a) Explain why.
b) How does it reach the roots?

11 List two differences between xylem cells and the cells that carry food (phloem cells).

Leaves

How is a leaf adapted for its function?

These Amazonian lily plants have huge leaves. Many flowering plants have leaves with a large **surface area** to help them trap as much light energy as possible from the Sun.

Veins carry substances into and out of the leaf. They contain **xylem tissue** and **phloem tissue**. The veins also help to support the leaf.

The **leaf stalk** helps to hold the leaf in the right place so that it can trap as much light energy as possible.

A

B

The function of the leaf is produce food for the plant using **photosynthesis**. Light energy is needed to make this process happen and so the more light energy a leaf gets, the more food it can produce.

The world's longest leaves belong to the Amazonian Palm plant. These leaves are nearly 20 m long. They are divided into smaller 'leaflets' supported by a large leaf stalk.

C

1 a) What is the name of the process that plants use to make food?
b) What energy does a plant need to make this process happen?
c) How do leaf stalks help the process?

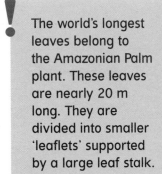

D *The leaf mosaic in a rhododendron plant.*

E *Beech trees have a very effective leaf mosaic.*

Most plants that grow in the UK have many leaves. These are often arranged so that they do not shade each other, and can all get light. This arrangement is called a **leaf mosaic**.

2 a) What is a leaf mosaic?
b) How does it help the plant?

3 Look at the beech wood in picture E. Why do you think there are very few plants growing beneath the trees?

Supporting the leaves

Leaf stalks hold the leaves in the right position to trap sunlight. The veins in a leaf also help to support it. However, water is also needed. When a plant does not have enough water it wilts. A plant **wilts** because the cells in the leaf and leaf stalk become 'floppy' if they start to dry up. They need water to keep their shapes.

4 a) What does a plant look like when it wilts?

b) Why does this happen?

The **cell sap** in the **vacuole** helps to keep the cell's shape by pushing outwards, a bit like a balloon.

  A well watered plant.

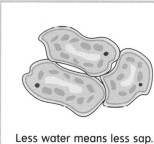

Less water means less sap.

  The plant has now wilted because it does not have enough water.

P How could you estimate (work out roughly) the total surface area of all the leaves on a plant?

F

Summary

A plant makes food using _____. This process needs _____ energy from the Sun and so leaves often have a large _____ _____ to trap this energy. The leaves have _____ which carry substances into and out of the leaf and also help to _____ the leaf. The leaves are also supported by having cells filled with water. If there is not enough water, the plant _____.

light   photosynthesis   support
surface area   veins   wilts

5 Why do most leaves have a large surface area?

6 a) Write down two ways in which a leaf is supported.

b) Why is it important that leaves are supported?

7 The largest vein in the leaf runs down its centre. This is called the mid-rib.

a) Draw a sketch of a leaf and label these parts: vein, mid-rib, stalk.

b) All these parts have three functions. What are they?

c) Write down the names of two tissues that are found in all of these parts.

Photosynthesis: raw materials

How does a plant make its own food?

Many people think that venus fly trap plants eat food just like humans. This is *not* true. These plants grow in areas where there are very few nitrates in the soil. Nitrates are very important mineral salts. The plants catch insects and use them as a source of nitrates, *not* food. Like all plants, venus fly traps **photosynthesise** to make food.

? **1** What process do venus fly trap plants use to make food?

A *A venus fly trap.*

B *The plants on the left are growing in soil without any nitrates. The plants on the right are growing in soil with nitrates.*

Most photosynthesis occurs in leaves. The top surface of a leaf is covered in **cuticle** which makes the leaf waterproof and stops it drying out too much. The cells in the upper and lower sides of the leaf form **epidermis tissue**. This tissue forms a 'skin' around the leaf, helping to hold it together.

? **2** Many desert plants have very thick layers of cuticle. Why do you think this is?

The main tissues in a leaf. **C**

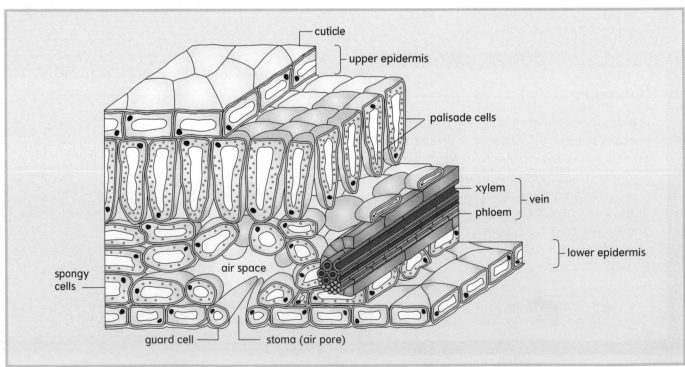

- cuticle
- upper epidermis
- palisade cells
- xylem
- vein
- phloem
- lower epidermis
- air space
- spongy cells
- guard cell
- stoma (air pore)

Photosynthesis mainly occurs in the palisade cells and needs two raw materials, **carbon dioxide** and **water**. The water comes from the roots. The carbon dioxide comes from the air and gets into the leaf through small holes called **stomata**. (The singular is **stoma**). Each stoma has two **guard cells** which open and close it.

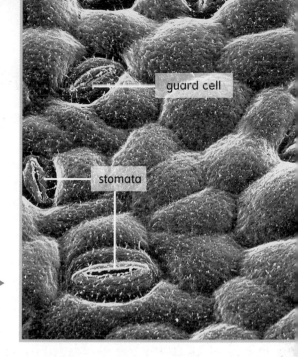

guard cell

stomata

P If leaves are put into very hot water, gases bubble out of the stomata.

- How would you show where stomata are found in different leaves?
- Are stomata found in different places depending on where a plant lives?

? **3** How does carbon dioxide get into the leaf?

Stomata are opened and closed by guard cells. **D**

E

The air spaces make sure that all the cells get a supply of carbon dioxide if they need it. Leaves are often thin, so that the carbon dioxide does not have to go very far before reaching cells that need it. The carbon dioxide gets into the leaves and cells by **diffusion**. This simply means that it moves from where there is a lot of it, to areas where there is less of it.

? **4** What is diffusion?

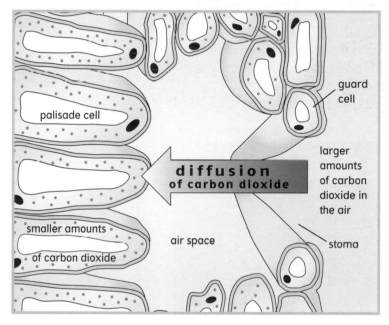

palisade cell

guard cell

diffusion of carbon dioxide

larger amounts of carbon dioxide in the air

smaller amounts of carbon dioxide

air space

stoma

Summary

Photosynthesis requires two raw materials: _____ _____ and _____ The _____ _____ gets into the leaf through small holes called _____. Each of these holes is surrounded by two _____ _____. Leaves are often _____ so that the _____ _____ does not need to travel very far before reaching a cell that needs it. Carbon dioxide gets into leaves and cells by _____.

carbon dioxide   diffusion
guard cells   stomata
thin   water

? **5** **a)** Why do many leaves have a large surface area?
b) Why are many leaves thin?

6 Which of these substances are the raw materials for photosynthesis?

oxygen   carbon dioxide   light energy
food   water

7 **a)** What are stomata?
b) What are they used for?
c) The Amazonian lily on page 54, lives in water. Where do you think its stomata are? Explain your answer.

B7 The transpiration stream

How do leaves get water for photosynthesis?

Water is very important for plants but many plants appear to waste a lot of it. When it is hot, about 700 litres of water will evaporate from the leaves of this oak tree each day.

The loss of water from leaves is called **transpiration** and it actually helps the plant because:

- it keeps the leaves cool
- it helps to suck up more water (containing mineral salts) from the roots. Mineral salts are needed to keep the plant healthy.

The passage of water from the roots, through the stem and out through the leaves is called the **transpiration stream**.

A *An oak tree.*

 1 a) What is transpiration?
b) How does transpiration help the plant?

How transpiration works

As water evaporates from the cells, the cells take more water out of nearby cells, which in turn take more water out of the xylem. This creates a 'sucking' action which pulls more water from the xylem further down the plant.

B

 These photographs show how guard cells change shape to open and close the stomata.

C

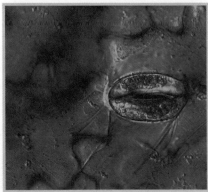

D

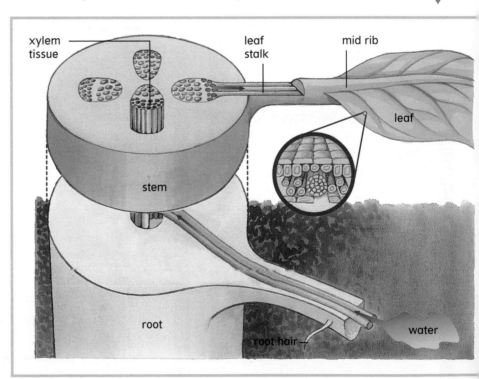

If the leaf is not getting enough water, the leaf tries to slow down transpiration by shutting the stomata. When there is plenty of water, the guard cells are full of water and the stomata are open. When there is not enough water, the guard cells become soft and the stomata close. This can help to stop a plant wilting.

2 a) Which photograph (C or D) shows cells from a leaf that is running out of water?

b) Explain your answer.

Factors that affect transpiration

Transpiration is faster when

- it is hot
- there is a lot of light
- it is windy
- there is not much water vapour in the air (low humidity).

3 Which weather conditions would make transpiration slower?

Summary

The loss of water from a plant is called _____. The route that water takes from the roots, through the _____ and out through the _____ in the leaves is called the _____ _____. This helps to keep the plant _____ and makes sure that _____ _____ are carried up the plant in the water. Transpiration is fastest when the _____ is high, the Sun is bright, it is _____ and when the air has a low _____.

cool	humidity	mineral salts
stem	stomata	temperature
	transpiration	windy
	transpiration stream	

How could you find out which weather conditions cause the greatest amount of water loss from a leaf?

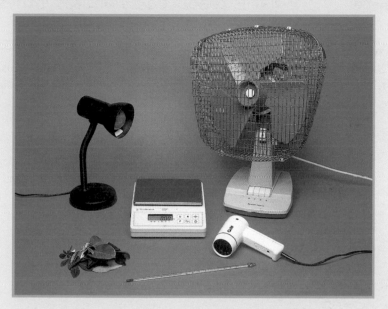

E

4 Graph F shows the mass of water lost by an oak tree during the course of two different days.

F

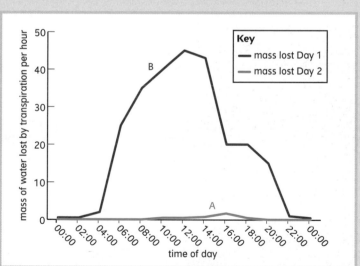

a) One set of results is from a day in summer and the other from a day in winter. Which is which?

b) Explain your answer to part a).

c) In the results for summer, what time do you think sunrise was?

d) Why do you think this?

5 Write a short description of the route taken by a water molecule in the transpiration stream.

Photosynthesis: products

What is produced by photosynthesis?

Animals would not exist if plants did not photosynthesise. The process that plants use to make their food also provides animals with food and oxygen. Photosynthesis is a **chemical reaction** and can be written down as a **word equation**.

carbon dioxide + water (+ light energy) ⟶ glucose + oxygen

reactants (raw materials)   products

1 What are the raw materials needed for photosynthesis?

2 What is the name of the substance that the plant makes for food?

The **reactants** are the **raw materials** needed for a chemical reaction. The **products** are the substances made by a chemical reaction. There are two more things that are needed to allow photosynthesis to happen: **chlorophyll** (a green substance found in **chloroplasts**) and **light energy** (normally from the Sun). The light energy goes in brackets in the word equation because it is not a substance, but it is needed to make the reaction happen.

Some plants do not have green leaves but they still contain chlorophyll. It's just that other chemicals inside the leaf disguise the green colour of the chlorophyll. A large beech tree, like this one, produces enough oxygen each day to keep four people alive.

3 Where do the reactants in photosynthesis come from?

The products of photosynthesis are **glucose** and **oxygen**. The oxygen **diffuses** out of the cells into the air spaces and out through the stomata. Since most of the oxygen is not needed it is often called a **by-product** of photosynthesis.

Some of the glucose is used by cells in **respiration**. Respiration is a chemical reaction which releases energy for the cell to use. Respiration happens in all living cells, all the time.

Glucose is a type of **soluble** substance called a sugar. A lot of the glucose is turned into **starch** and stored. Starch is **insoluble**. Starch is a better way to store food since it will not dissolve and get out of the chloroplast — it will stay where it is put!

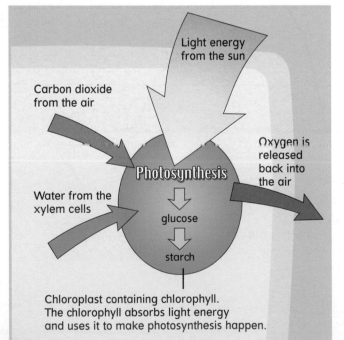

Light energy from the sun

Carbon dioxide from the air

Oxygen is released back into the air

Photosynthesis

Water from the xylem cells

glucose

starch

Chloroplast containing chlorophyll. The chlorophyll absorbs light energy and uses it to make photosynthesis happen.

B

4 Why does the plant change glucose to starch for storage?

P If you boil a leaf in water for a minute and then leave it in hot ethanol for 10 minutes, the chlorophyll is removed. You can test for starch in this leaf using iodine solution, which turns starch blue-black. How would you show that light and chlorophyll are needed for photosynthesis?

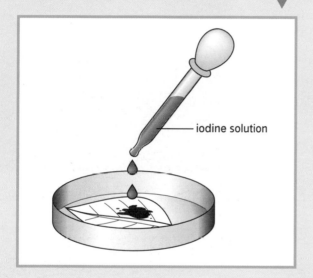

C

iodine solution

5 **a)** What gas is a by-product of photosynthesis?
b) How does it leave the leaf?

6 Look at the plant in picture D. For each lettered leaf, write down what colour you think it will turn when it is tested for starch. Explain each answer.

D

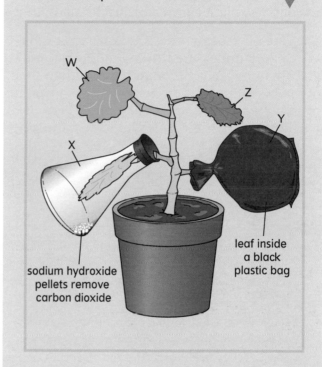

W

Z

Y

X

sodium hydroxide pellets remove carbon dioxide

leaf inside a black plastic bag

Summary

The word equation for photosynthesis is:

_____ _____ + _____(+ _____energy) ⟶ _____ + _____

The reactants are _____ _____and _____. Photosynthesis also needs _____ _____to make it happen. The products are _____ and _____. Some of the glucose is used for _____. This is a chemical reaction that releases _____for a cell. A lot of the glucose is turned into _____and stored. Starch is a useful storage material because it is _____. It can be tested for by using _____solution which turns it blue/black.

carbon dioxide   energy   glucose   insoluble   iodine   light
light energy   oxygen   respiration   starch   water

7 Plants need some oxygen for respiration. Plants that live underwater put the oxygen they produce into the water. They may also take oxygen out of the water.

Do you think the amount of oxygen in a river is highest during the day or at night? Explain your answer fully.

Rate of photosynthesis

What factors control the rate of photosynthesis?

In high mountain regions the air is 'thinner', which means that there are fewer air molecules. This means there is less carbon dioxide for photosynthesis. It is also colder. For these reasons, the plants grow slowly and many stay quite small.

The **rate** of photosynthesis is how fast it is happening. We can often measure this by finding out how much glucose or oxygen is produced in a certain time. The rate of photosynthesis is affected by the temperature, the amount of carbon dioxide and the amount of light. If there is not much of any of these, it is said to be a **limiting factor**.

1   What are the three limiting factors in photosynthesis?

P

Cress leaves that have been cut in half and placed in a syringe will float when they are given light. This is because photosynthesis makes bubbles of oxygen which stick to the leaves.

How would you use this apparatus to compare how quickly photosynthesis happens when the cress leaves are given different strengths of light?

B

A   *These plants are growing in the mountains.*

The oxygen produced by most underwater plants bubbles into the water. However, some duckweeds form balls of weed in which the oxygen collects. Every so often the ball of weed rises to the surface to get rid of the oxygen and then sinks again.

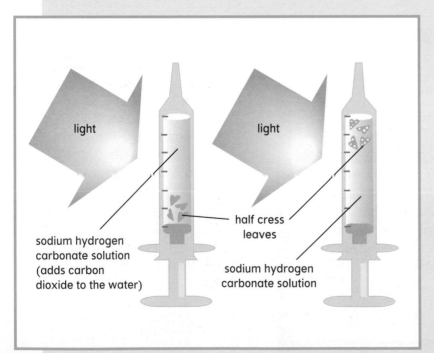

light

light

sodium hydrogen carbonate solution (adds carbon dioxide to the water)

half cress leaves

sodium hydrogen carbonate solution

C

Graphs can show how some factors affect the rate of photosynthesis.

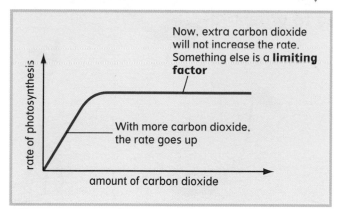

Now, extra carbon dioxide will not increase the rate. Something else is a **limiting factor**

With more carbon dioxide, the rate goes up

rate of photosynthesis

amount of carbon dioxide

Summary

The rate of photosynthesis is how _____ photosynthesis is happening. The rate is affected by a range of factors: the _____, the amount of _____ _____and the amount of _____. When one of these factors stops the rate of photosynthesis increasing it is said to be a _____ _____.

| carbon dioxide | fast | light |
| limiting factor | temperature | |

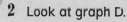 **2** Look at graph D.
a) Explain what a limiting factor is.
b) Where the graph becomes level, something has become a limiting factor. Which of these could it be? (There are more than one.)

temperature   nitrogen   oxygen   sugar
carbon dioxide   glucose   light

A similar graph can be drawn for temperature. You can see in graph E that if the temperature gets too hot, photosynthesis stops completely. This is because high temperatures (above about 45 °C) destroy the chlorophyll.

E

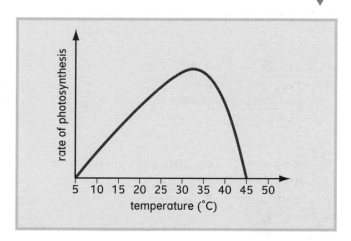

rate of photosynthesis

5 10 15 20 25 30 35 40 45 50
temperature (°C)

 3 a) Where is chlorophyll found?
b) What does it do?

4 In an experiment, the number of bubbles coming from some Canadian pondweed was counted each minute. The distance between a lamp and the pondweed was altered. Table G shows the results.

a) Plot these results on a line graph.
b) What is being used as a measure of the rate of photosynthesis?
c) What does your graph show (i.e. what is the relationship between the distance of the lamp and the number of bubbles)?
d) What was the maximum number of bubbles per minute?
e) How might the maximum number of bubbles be increased further?
f) At the end of the experiment, the bright lamp was left next to the pondweed by mistake. An hour later it was discovered that there were hardly any bubbles coming from the pondweed. Why do you think this might be so?

F

Distance between lamp and pondweed (cm)	Number of bubbles per minute
1	65
5	65
10	63
15	59
20	54
25	47
30	35

5 Why do you think plants grow slowly in high mountain regions? Use the words photosynthesis and limiting factor in your answer.

Plant growth

How do plants know which way to grow?

When a new plant starts to grow, its shoot always grows upwards and its roots always grow downwards. It does not matter which way up the seed was planted.

Chemicals called **plant hormones** help the plant do this. **Auxin** is an example of a **plant hormone**. It is found in the tips of shoots and causes growth. Auxin is only found in the tip of a shoot. If the shoot tip is cut off, the shoot will not grow any further.

If a shoot grows sideways, the auxin is pulled by gravity to the underside of the shoot. It causes this part of the shoot to grow faster than the top side and so the shoot grows upwards.

A

 1 a) What effect does auxin have in shoots?
   b) Why is this useful for the plant?

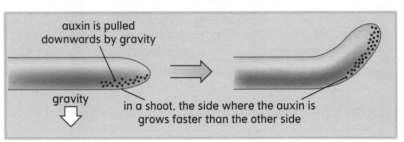

auxin is pulled downwards by gravity

gravity

in a shoot, the side where the auxin is grows faster than the other side

B   C

Auxin is also found in root tips where it has the *opposite effect*. The auxin slows down growth in the growing root.

 2 a) What effect does auxin have in roots?
   b) Why is this useful for the plant?

gravity

in a root, the side where the auxin is grows slower than the other side

D

Plant shoots also grow towards the light. This is also due to the effects of auxin. The auxin moves to the part of the shoot tip in the shade. This again makes that part of the shoot grow faster.

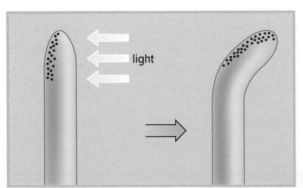

light

E

Moisture in the soil causes auxin to move to the part of the root nearest the moisture. Again, in the root auxin slows down growth on that side of the root causing the root to grow towards the moisture.

3 Why do you think it is important that roots grow towards moisture?

P How would you find out whether shoots can still move towards light if the tip of the shoot is either covered in kitchen foil or cut off?

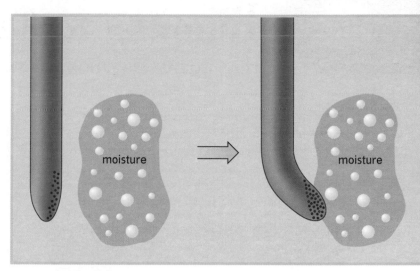

4 Draw what you think would happen to the root and shoot in each of the experiments shown in picture G. Each drawing shows what the plant looked like at the start of the experiment.

G

a) light ↓↓↓ soil surface
shoot   root

b) light ↓↓↓

c) light soil surface

d) light soil surface

e) light   shoot tip removed   soil surface   moisture

f) light   blob of vaseline (containing auxin)   soil surface

! Charles Darwin, the scientist who proposed the theory of evolution, was the first scientist to investigate why shoots grow towards the light, in 1880.

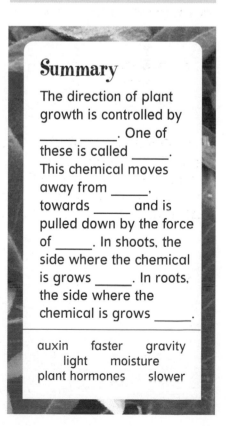

Summary

The direction of plant growth is controlled by _____ _____. One of these is called _____. This chemical moves away from _____, towards _____ and is pulled down by the force of _____. In shoots, the side where the chemical is grows _____. In roots, the side where the chemical is grows _____.

auxin   faster   gravity
light   moisture
plant hormones   slower

B11 Humans and plant hormones

How do humans make use of plant hormones?

Many people prefer to eat seedless grapes because they do not contain pips. Some seedless fruits occur naturally but others are produced using plant hormones. The flowers are sprayed with a plant hormone which causes the fruits to form but not the seeds.

 **A** *Grapes growing in Spain.*

 B

Grape plants are sprayed with a plant hormone to make seedless fruits.

Fruit ripening is also controlled using plant hormones. Plant hormones are sprayed onto:

- fruit trees to stop the fruit falling off. Bigger fruits are produced.
- fruit trees to speed up ripening. All the fruit ripens together and can be picked in one go.
- unripe fruit to make them ripe. The fruit will reach the supermarkets in a perfect, 'just ripened' condition.

C *Bananas are picked unripe in the Caribbean and transported by ship to the UK. They are then ripened using plant hormones before going to the supermarket.*

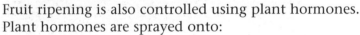

1 What do you think would happen if bananas were picked ripe in the Caribbean, which is over 9000 kilometres from the UK?

Large numbers of plants can be produced quickly using rooting powders. These contain different plant hormones which makes the roots grow more quickly. Parts of a plant (**cuttings**) are taken, dipped in rooting powder and placed in water. After the roots have grown, the cuttings are planted and develop into new plants.

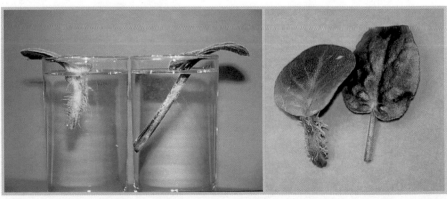

 D *The cutting on the left was dipped in rooting powder.*

? **2** Many plant cuttings develop roots without using rooting powder. What is the advantage of using rooting powder on cuttings?

P How would you find out whether some makes of rooting powder are better than others?

! Cut flowers produce a plant hormone which makes the flowers die faster. Adding aspirin to their water stops the production of this plant hormone so the flowers last longer. Sometimes, cut flowers are sold with a packet of a chemical which does the same thing. **E**

Plant hormones are also used as weedkillers. The plant hormones change the way plants with broad leaves grow and so they die. Plants with narrow leaves are not affected and so farmers can kill all the weeds in a field of a cereal crop (like wheat) without affecting their crop.

F *This spray, containing plant hormones, will kill the weeds but not the wheat.*

? **3** What is the advantage of using a weedkiller containing plant hormones over one that kills all plants?

! Agent Orange was a weedkiller containing plant hormones used in the Vietnam War. It destroyed the jungle so that enemy movements could be seen by the Americans. It also caused many health problems to people who came into contact with it.

? **4** Weeds are a problem on playing fields. What do you think is the easiest way of getting rid of the weeds on a playing field, without destroying the grass? Explain your answer.

5 There is a huge range of exotic fruit in supermarkets. 25 years ago the range was much smaller. Explain why you think this is so.

Summary

Plant _____ are used by humans to control plants. Large numbers of plants can be produced _____ using _____ _____ which makes roots grow from cuttings quickly. The ripening of _____ can also be controlled so that it looks perfect in supermarkets. Weeds can be killed too. Plant hormones are used to kill plants with _____ leaves but they do not affect plants with _____ leaves.

| broad | fruit | hormones | narrow |
| quickly | rooting powder | |

Human senses

How do humans sense things?

Most people think that humans only have five senses — sight, hearing, taste, touch and smell. In fact we have six. We also have a sense of balance.

Many of our organs and tissues contain **receptors** which detect changes in our surroundings. These changes are called **stimuli** (the singular is **stimulus**).

All the information that the receptors receive is sent to the brain along **nerves**.

A B

1 Draw a table like this to show which sense organs are used to detect which stimuli.

Sense	Organ where this sense is detected	Stimulus that is detected

The eye contains receptors that sense light. These receptors are found at the back of the eye in the **retina**. Information from these receptors is passed to the brain along the **optic nerve**.

2 How does information get from the eye to the brain?

C

! If you hold your nose or have a cold, you cannot tell the difference between the tastes of some things like tea and coffee. This is because your sense of smell is also involved in tasting things.

The ear contains receptors which are sensitive to sound. There are also receptors that sense changes in your position to help you keep your balance.

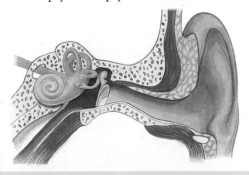

3 What stimuli do the ears detect?

D

Receptors in the roof of the nose are sensitive to chemicals in the air.

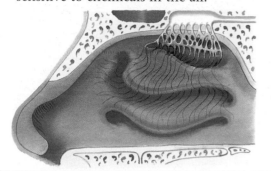

 4 Why do you think people lose their sense of smell when they have a cold?

 E

Your skin is also a sense organ. It contains receptors called **nerve endings** which detect things like pressure and temperature.

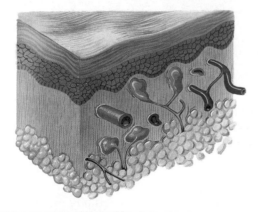

 7 There are two sorts of pressure receptors in the skin. Some for heavy pressure and others for light pressure. Which type do you think are closer to the surface of the skin?

G

P Some parts of your skin are more sensitive than other parts. If two pencils are held together and pressed gently onto the skin, some areas of your skin will feel two points and others only one. How would you find out which parts of the skin are the most sensitive?

Receptors in the tongue, called **taste buds**, sense chemicals in our food and allow us to taste things. You can only taste chemicals which dissolve in the liquid in your mouth.

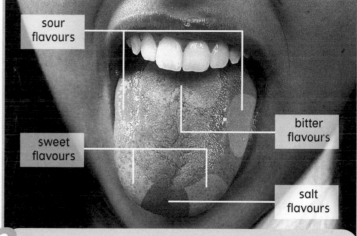

sour flavours

sweet flavours

bitter flavours

salt flavours

5 Where on the tongue do you think receptors would pick up the taste of a chocolate bar?

F

6 Why can't you taste insoluble things?

Summary

Some organs contain _____ which detect changes in the surroundings. These changes are called _____. Receptors in the eye are sensitive to _____; those in the _____ are sensitive to sound and also to changes in body _____; those in the tongue and nose are sensitive to _____ in our food or in the _____; and those in the skin are sensitive to pressure and changes in _____. Information from the receptors goes to the _____ along _____.

air   brain   chemicals   ear   light   nerves
    position   receptors   stimuli   temperature

The eye

How do humans see things?

When a photograph is taken, the picture is actually upside down. You can see this by using a pin-hole camera with a screen at the back.

A

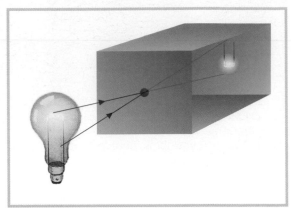

B

Inside the eye

The **sclera** is a tough layer surrounding the eye which protects it. The part at the front is called the **cornea**, which is transparent so that light can get into the eye. The cornea is curved which bends light rays helping to make a clear image on the retina.

The **pupil** is a gap in the middle of the **iris** that light goes through. The iris is a muscle which makes the pupil larger or smaller.

The **cornea** and the **lens** both bend light rays to make the image on the retina clear (in focus). Many people's lenses do not work very well and they wear extra lenses (glasses).

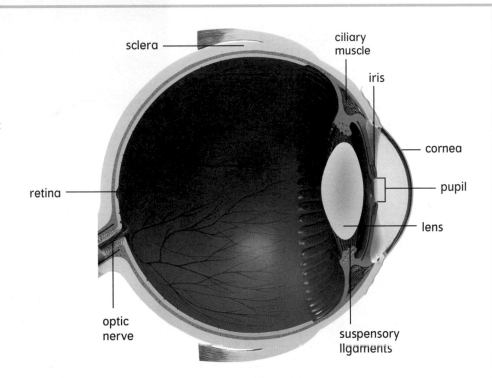

sclera · ciliary muscle · iris · cornea · pupil · lens · suspensory ligaments · retina · optic nerve

The lens is held in place by the **suspensory ligaments** and the **ciliary muscles**.

The **retina** contains the receptors that are sensitive to light.

The **optic nerve** is actually many nerves bundled together, taking information from all over the retina to the brain. The brain turns this information into what you see, which is the right way up.

1 a) Which two parts of the eye bend light rays to make the image on the retina in focus?
b) Why do many people need to wear glasses?

C

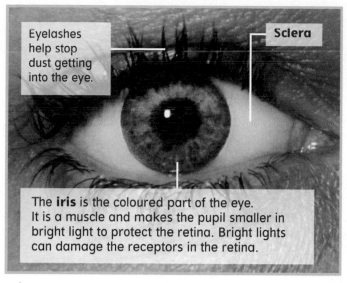

Eyelashes help stop dust getting into the eye.

Sclera

The **iris** is the coloured part of the eye. It is a muscle and makes the pupil smaller in bright light to protect the retina. Bright lights can damage the receptors in the retina.

 D

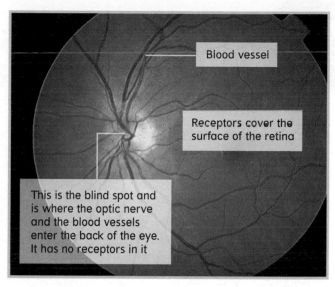

Blood vessel

Receptors cover the surface of the retina

This is the blind spot and is where the optic nerve and the blood vessels enter the back of the eye. It has no receptors in it

 E *The author's retina.*

? **2** In a dark room, will the pupil be large or small?

? **3** Why do you think the blind spot is called that?

P You can find your blind spot by looking at these two apples. Close your left eye and look at the red apple. Move the book towards you. What do you notice? How far away from your face was the book when this happened? Did this happen for everyone in the class at the same distance?

F

? **4 a)** What part of your eye contains the colour?
   b) What is the white part of the eye called?

5 How are the eyes protected from dust?

6 Which two parts of your eye are transparent?

7 a) What happens to your eyes when you walk into bright sunshine?
   b) Why do you think it is a good idea that this happens?
   c) Why do you think you should never look directly at the Sun?

Summary

Part of eye	What it does
	Contains light receptors.
	Focuses the image on the retina.
	Gap in the middle of the iris.
	Is attached to the suspensory ligaments and helps to support the lens.
	Muscle that opens and closes the pupil.
	Part of the sclera that lets light through. It also focuses images onto the retina.
	Protects the whole of the eye.
	Attached to and supports the lens.
	Takes information from the receptors back to the brain.

The nervous system

How do humans detect and respond to stimuli?

Information also needs to be sorted and directed around your body and this is done by the nervous system. The **nervous system** contains many **nerve cells** which are connected together so that they can pass information between each other. The nervous system also contains two organs.

A nerve cell or neurone.

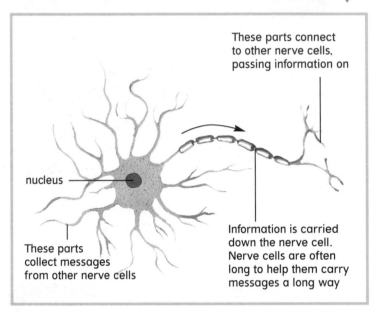

These parts connect to other nerve cells, passing information on

nucleus

These parts collect messages from other nerve cells

Information is carried down the nerve cell. Nerve cells are often long to help them carry messages a long way

1 a) What is the function of a nerve cell?
   b) How is a nerve cell adapted to this function?

2 Name the two organs found in the nervous system.

Receptors detect changes in and around our bodies and send information back to the brain. This information is sent as electrical signals called **impulses** which travel along nerve cells called sensory neurones. For example, the eye contains receptors. It sends impulses down a bundle of **sensory neurones** called the **optic nerve** to the brain. Impulses from receptors below the neck travel through the **spinal cord** up to the brain.

Normally your brain decides what to do when it receives an impulse. It is said to **process** the information that it receives. It can then send impulses telling part of your body to do something. These parts of the body are called **effectors**. For example, muscles are effectors and move when they receive impulses.

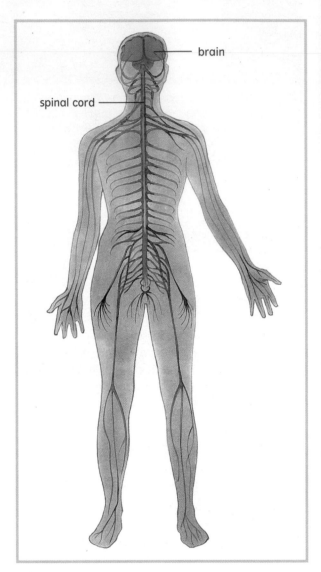

brain

spinal cord

A *The nervous system contains nerves and two organs.*

3 What route would an impulse take from a receptor in the skin of your foot to your brain?

Receptors detect changes called **stimuli**. The brain tells parts of your body what to do when a stimulus is detected. This is called the **response**. The body is said to respond to stimuli. In this way your brain co-ordinates all your actions.

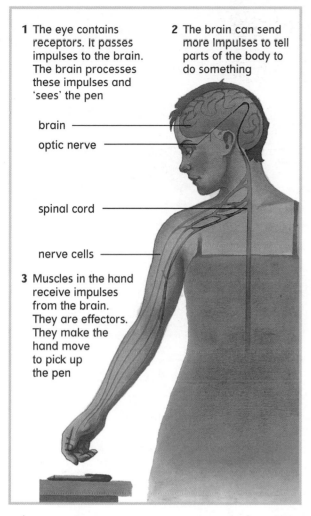

1 The eye contains receptors. It passes impulses to the brain. The brain processes these impulses and 'sees' the pen

brain

optic nerve

spinal cord

nerve cells

2 The brain can send more impulses to tell parts of the body to do something

3 Muscles in the hand receive impulses from the brain. They are effectors. They make the hand move to pick up the pen

D *This is what happens in the nervous system when you pick up a pen.*

Sometimes it is important to respond to a stimulus as fast as possible. For example, removing your hand from a hot object. An automatic response like this is called a **reflex action**. The brain does not do any processing to make this response happen, since this would slow the response down. Instead, the impulse passes along a sensory neurone to the brain or spinal cord and then goes straight to a motor neurone. In the case of removing your hand from a hot object, a muscle in the arm is the effector. Glands (eg. salivary glands and glands in the pancreas) can also be effectors and release a secretion when they receive an impulse.

E

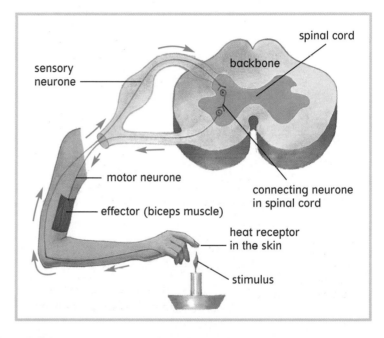

spinal cord

backbone

sensory neurone

motor neurone

effector (biceps muscle)

connecting neurone in spinal cord

heat receptor in the skin

stimulus

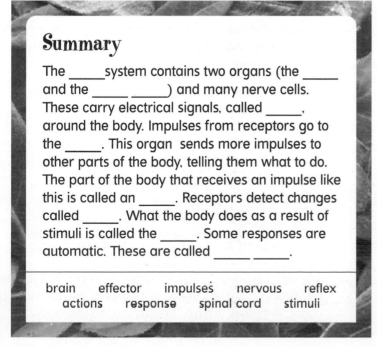

Summary

The _____ system contains two organs (the _____ and the _____ _____) and many nerve cells. These carry electrical signals, called _____, around the body. Impulses from receptors go to the _____. This organ sends more impulses to other parts of the body, telling them what to do. The part of the body that receives an impulse like this is called an _____. Receptors detect changes called _____. What the body does as a result of stimuli is called the _____. Some responses are automatic. These are called _____ _____.

   brain   effector   impulses   nervous   reflex
    actions   response   spinal cord   stimuli

4 a) What is a reflex action?
  b) Why is a reflex action quicker than a normal response?

5 A song you like is played on the radio and you turn up the volume. Explain what happens in the nervous system when you do this.

6 Explain what is meant by a response to a stimulus.

7 a) Why do nerve cells need to connect to each other?
  b) Draw two nerve cells connected together.
  c) In what way is the internet similar to the brain?

Excretion

How are waste products removed from your body?

Waste needs to be cleared away. Waste that is not cleared away starts to rot and can become very unpleasant. Your body also produces waste which needs to be removed.

Many chemical reactions which happen inside your body produce poisonous waste that needs to be removed. Removing these wastes is called **excretion**.

1 **a)** Why does your body need to remove the waste products from chemical reactions?
   b) What is this process called?

A

Respiration is a chemical reaction that occurs in all your living cells and releases energy from glucose (a type of sugar).

The carbon dioxide needs to be removed. It leaves the cells where it has been made and dissolves in the blood and is then carried to the lungs. It is removed from your blood in the lungs and leaves your body when you breathe out.

2 **a)** What process produces carbon dioxide?
   b) Where does this process occur?
   c) Which organs get rid of the carbon dioxide?

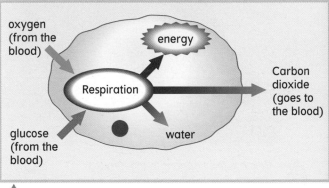

oxygen (from the blood)
energy
Respiration
glucose (from the blood)
water
Carbon dioxide (goes to the blood)

B *Respiration*

As well as carbon dioxide, your blood contains many other poisonous waste products that need to be removed.

The food that you eat contains **proteins**. These are broken down (**digested**) in the **digestive system** into **amino acids**. Amino acids are very important for your body but if you have more than you need, the liver breaks them down into a substance called **urea**.

3 What does the liver do to amino acids that are not needed?

The urea goes into the blood and is removed from the blood by the **kidneys**. The urea is dissolved in water, forming **urine**. The urine is stored in the **bladder** until you go to the toilet and **urinate**.

P

How would you show that there is more carbon dioxide in the air that you breathe out than the air you breathe in?

C

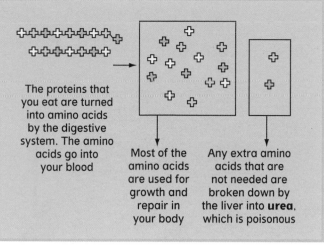

The proteins that you eat are turned into amino acids by the digestive system. The amino acids go into your blood

Most of the amino acids are used for growth and repair in your body

Any extra amino acids that are not needed are broken down by the liver into **urea**, which is poisonous

D

! All your blood passes through your kidneys, to be cleaned, every five minutes.

4 a) What substance do the kidneys excrete?
b) What substance do the lungs excrete?

5 The percentage of carbon dioxide in two samples of air was measured. Sample A contained 0.04 % and sample B contained 4.1 %. Which sample do you think was 'breathed out' air? Explain your reasoning.

6 The person in the photograph has kidneys that do not work. Her blood supply is flowing through a dialysis machine.
a) What do you think the dialysis machine is doing?
b) Why does the machine need to do this?

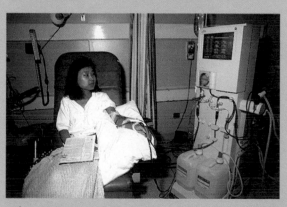

F

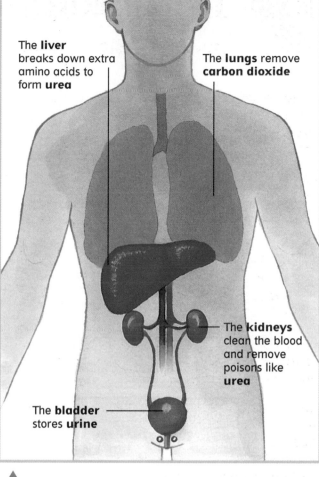

The **liver** breaks down extra amino acids to form **urea**

The **lungs** remove **carbon dioxide**

The **kidneys** clean the blood and remove poisons like **urea**

The **bladder** stores **urine**

E

Summary

Poisonous substances made by chemical _____ in the body need to be removed. This is called _____. _____ _____ is a gas produced by a process that releases energy in cells, called _____. This gas is removed by the _____. The _____ takes amino acids that are not needed and breaks them down to form _____. This substances is removed by the _____. It is dissolved in water to form a liquid, called _____, which is stored in the _____ until you urinate.

bladder   carbon dioxide   excretion
kidneys   liver   lungs   reactions
respiration   urea   urine

Homeostasis

How does your body keep your internal conditions constant?

Control of temperature

A thermostat is used in homes to keep a constant temperature. If the house gets colder than a set temperature, it turns the heating on. When the right temperature is reached, it switches the heating off again.

The chemical reactions in your body need substances called **enzymes**, which work best at 37 °C.

1 Graph B shows how well an enzyme works at different temperatures. Which letter represents a temperature of 37 °C?

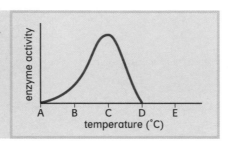

B

Your brain acts a bit like a thermostat. If your body temperature goes above 37 °C, your brain detects this and sends impulses down nerve cells to the skin to make you sweat. Sweat evaporating from the skin cools you down.

If your body temperature goes below 37 °C, impulses are sent to muscles which make you shiver. Shivering warms you up.

2 Which organ acts like a thermostat in your body?

Control of water content

You lose water when you breathe. The moisture on the surfaces of the lungs evaporates and so you breathe out water vapour.

dry

wet

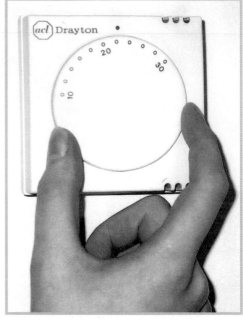

A *A thermostat.*

C *On a cold day, the water vapour in your breath condenses into clouds of water droplets that you can see.*

3 What colour does dry cobalt chloride paper change to when water is added?

D *Dry cobalt chloride paper changes colour when water is added to it.*

You also lose water when you sweat. If it is hot you sweat a lot and so you need to make sure that you drink plenty of water to replace the loss. Water is also lost in urine.

If your body has more water than it needs, the extra water is lost (through the kidneys) in your urine. If your body has less water than it needs, your brain makes you become thirsty.

4 Write down three ways in which your body loses water.

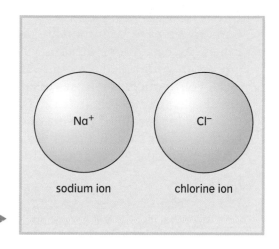

sodium ion   chlorine ion

Sodium chloride (common salt) is made of sodium ions and chloride ions. Sodium ions are important for allowing nerve cells to carry impulses. **F**

Control of ion content

Ions are atoms or molecules that carry an electrical charge. Some ions are important for the body.

You lose ions in your sweat. Any extra ions are removed from the body by the kidneys.

Keeping all the conditions inside your body constant is called **homeostasis**.

Most people need to get about 2 litres of water into their bodies each day, from food and drink. Visitors to very hot places must drink between 4 and 8 litres each day.

Death Valley in California, the hottest place on Earth. **E**

Summary

Your body controls the conditions inside it and keeps them constant. This is called _____. Three conditions that are controlled include _____, the amount of _____ and the amount of _____. Your body is kept at about 37 °C because human _____ work best at that temperature. You lose water from your _____ when you breathe out, from your _____ when you sweat and in your _____. Ions are lost from your skin when you _____. Any extra water or ions, that are not needed by your body, are removed by your _____ and put into your _____.

| enzymes | homeostasis | ions | kidneys | lungs |
| skin | sweat | temperature | urine | water |

5 a) What is homeostasis?
b) Write down three conditions that the body controls.

6 How are the following lost from your body?

a) water
b) ions
c) heat.

7 In very hot places, some people take salt tablets. Why do you think they do this?

Hormones

How do hormones control processes in your body?

This person has a disease called **diabetes**. Many **diabetics** have to inject themselves with a substance called **insulin**.

Carbohydrates in your food are mainly digested into a sugar called **glucose**. After a meal the amount (concentration) of glucose in your blood goes up. When it gets above a certain level, your **pancreas** releases **insulin**. The pancreas contains gland cells that **secrete** (release) the insulin. The insulin dissolves in the **blood plasma** and is carried around your body in your blood. It affects certain cells in your liver, which then take glucose out of the blood, and store it.

 1 **a)** Where is insulin produced?
b) How does it travel around the body?

Diabetics cannot produce their own insulin and so their blood glucose levels may get too high. This can cause severe damage to the kidneys and brain and cause death. To stop this happening they can:

- make sure that they do not eat too many carbohydrates
- inject themselves with insulin.

A *This is a daily routine for diabetics.*

2 What organs are damaged if the glucose level in the blood gets too high?

When the glucose levels in the blood of a diabetic get too high, some of the glucose comes out in the urine. Doctors test this with glucose test strips. The ancient Romans used to put saucers of urine out in the Sun. If bees visited the urine, they knew the person was diabetic.

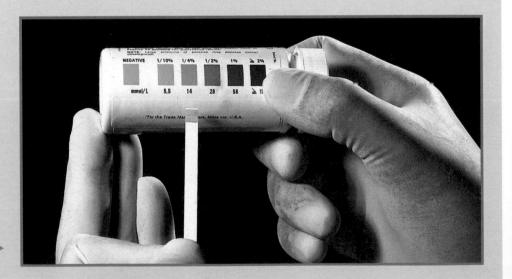

B

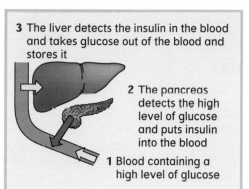

3 The liver detects the insulin in the blood and takes glucose out of the blood and stores it

2 The pancreas detects the high level of glucose and puts insulin into the blood

1 Blood containing a high level of glucose

C

D

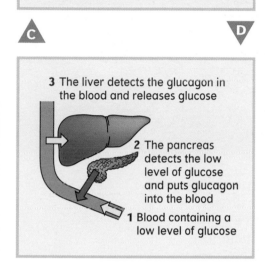

3 The liver detects the glucagon in the blood and releases glucose

2 The pancreas detects the low level of glucose and puts glucagon into the blood

1 Blood containing a low level of glucose

When your blood glucose concentration falls below a certain level, your pancreas secretes another substance, called **glucagon**. This is also carried in the blood and causes the cells in the liver to release glucose back into the blood.

?

3 a) This is another example of **homeostasis**. What does this word mean?

b) The levels of what substance are being controlled?

Insulin and glucagon are **hormones**. Hormones are 'chemical messengers', making various processes in the body happen.

All hormones are carried in the liquid part of the blood (the **plasma**) and are produced by parts of the body called **glands**. The pancreas contains glands. The organs or cells that affect hormones are called **target organs** or **target cells**.

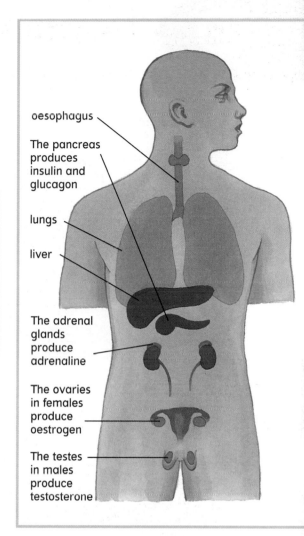

oesophagus

The pancreas produces insulin and glucagon

lungs

liver

The adrenal glands produce adrenaline

The ovaries in females produce oestrogen

The testes in males produce testosterone

E *The human body produces many different hormones.*

?

4 a) What is a hormone?
b) What parts of the body produce hormones?

5 Write down two ways that diabetics use to control the concentration of glucose in their blood.

6 Nerve cells and hormones both carry 'messages' around the body. How fast do you think messages are carried by hormones, compared with nerve cells? Explain your reasoning.

7 Find out what the other hormones labelled in picture E do.

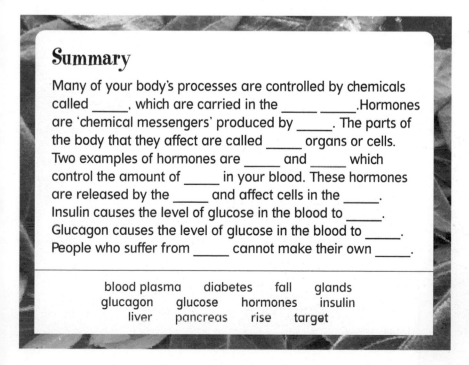

Summary

Many of your body's processes are controlled by chemicals called _____, which are carried in the _____ _____. Hormones are 'chemical messengers' produced by _____. The parts of the body that they affect are called _____ organs or cells. Two examples of hormones are _____ and _____ which control the amount of _____ in your blood. These hormones are released by the _____ and affect cells in the _____. Insulin causes the level of glucose in the blood to _____. Glucagon causes the level of glucose in the blood to _____. People who suffer from _____ cannot make their own _____.

blood plasma   diabetes   fall   glands
glucagon   glucose   hormones   insulin
liver   pancreas   rise   target

Smoking

How does smoking affect the human body?

In 1493, Spanish sailors returning from America, reported that the native people used to 'drink smoke'. This was Europe's first contact with **tobacco**. The leaves of the tobacco plant contain a **drug** called **nicotine**. People use tobacco to get this drug into their bodies in different ways.

A tobacco plant and some of its products.

Drugs are substances that affect the chemical reactions happening inside your body. Some drugs, called **medicines**, are useful and are used to treat diseases. Other drugs are not useful and are harmful. Many drugs are **addictive** (including some medicines) and people feel that they have to take them just to survive. A person like this is said to be **dependent** on a drug. Without the drug, a person may start to feel unwell and have **withdrawal symptoms**.

1　**a)** What is a drug?
　　b) What is the addictive drug found in tobacco ?
　　c) What is a medicine?

Nicotine is a poison. It is used in some sprays to kill insects. This product is so poisonous you cannot buy it in the shops.

Nicotine is a poison. In fact, it is used in to kill off insects. It can cause the heart to stop beating regularly and causes arteries to get narrower.

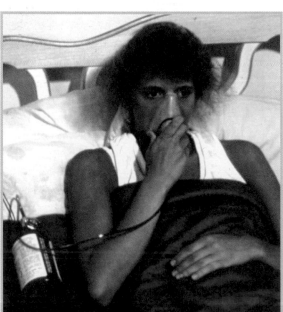

Narrow arteries cannot carry as much blood as they should and so less oxygen gets to parts of the body. When this happens to arteries supplying the heart muscle, the cells do not get enough oxygen and they die. This is called **heart disease**.

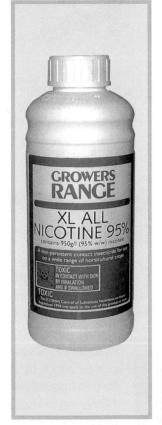

GROWERS RANGE

XL ALL NICOTINE 95%
(contains 950g/l (95% w/w) nicotine

A non-persistent contact insecticide for use on a wide range of horticultural crops

TOXIC
IN CONTACT WITH SKIN
BY INHALATION
AND IF SWALLOWED

TOXIC

Smoking has damaged this woman's lungs so much that she needs to carry oxygen around with her.

2 What are the harmful effects of nicotine?

Tobacco smoke also contains tar. Tar contains chemicals that cause **cancer**. Tar also irritates the lungs and causes more of a sticky liquid called **mucus** to be produced.

 3 What do the cilia in the lungs do?

The mucus sitting around in the lungs can get infected with bacteria and become very sore, causing **bronchitis**. Smokers cough to get the mucus out of their lungs. Coughing, over a long period of time, causes the delicate walls of the lungs to break apart and become swollen. This results in a disease called **emphysema**, in which people are permanently breathless.

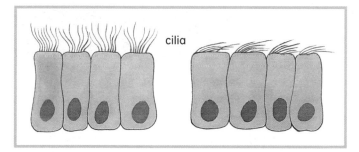

E *Normally, mucus is swept out of the lungs by **ciliated epithelial cells** which have hair-like **cilia** on them. However, the tar stops the cilia working.*

cancer tumour

D *The lung on the right is from a smoker. The tar makes the lungs go black and can cause cancer. 90 % of lung cancers are caused by smoking.*

Smoke also contains **carbon monoxide**, a poisonous gas which sticks to haemoglobin in red blood cells. Haemoglobin is a substance that carries oxygen around the body but if carbon monoxide sticks to it, oxygen cannot be carried. If a pregnant woman smokes, the growing baby (**fetus**) may not get enough oxygen and may be born too early and smaller than it should be.

Summary

_____ are chemicals which affect the chemical reactions happening in the body. Some are used to treat diseases and these are called _____. People may become _____ on a drug and suffer _____ symptoms without it. The addictive chemical in tobacco is called ___. This can cause _____ disease. Tar in smoke causes _____ and other diseases like bronchitis and _____ (when the lung walls break apart and swell up). _____ _____, a poisonous gas found in tobacco smoke, stops the red blood cells from carrying _____.

cancer   carbon monoxide   dependent
drugs   emphysema   heart   medicines
nicotine   oxygen   withdrawal

4 Make a list of all the diseases on these two pages and write down how each is caused.

5 Look at graph F. Why do you think people gradually realised that smoking causes cancer?

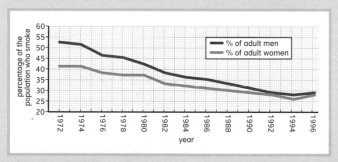

F *Percentage of adults in the UK who smoke.*

6 Many people die each year due to carbon monoxide produced by badly fitted or badly looked after gas central heating boilers. Carbon monoxide is a gas without any smell. Explain why the people die.

Other drugs

What are the affects of alcohol and solvents?

Drugs that slow down the nervous system are called **depressants**. Alcohol (**ethanol**) is an example. In small quantities it makes people feel good but it also slows down the time it takes for people to react to things.

In larger amounts alcohol causes vomiting. In very large amounts it can cause death since it stops the brain sending impulses to the lungs and so breathing stops.

 1 What is a depressant?

2 Why do you think people are advised not to drink and drive?

 Every week in the United Kingdom, 10 people are killed as a direct result of drinking and driving.

A *The effects of alcohol on the body.*

 3 Look at pictures A and B. Someone at a party drinks 2 pints of beer and 1 measure of vodka.

a) How many units will they have drunk?

b) What effect will this have on the person's body?

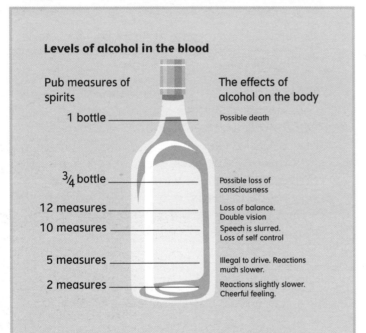

Levels of alcohol in the blood

Pub measures of spirits

	The effects of alcohol on the body
1 bottle	Possible death
¾ bottle	Possible loss of consciousness
12 measures	Loss of balance. Double vision
10 measures	Speech is slurred. Loss of self control
5 measures	Illegal to drive. Reactions much slower.
2 measures	Reactions slightly slower. Cheerful feeling.

The liver is responsible for destroying alcohol in the body. Drinking large amounts over a long time can damage the liver and can also damage the brain.

Like other drugs, alcohol is addictive. People who are dependent on alcohol are called **alcoholics**.

B

The amount of alcohol in different drinks varies. All these drinks contain the same amount of alcohol - 1 unit.

 ¼ a pint of strong beer lager or cider

 1 single pub measure of spirits e.g. vodka or whisky

 1 small glass of wine

 ½ a pint of ordinary beer, lager or cider

Solvent abuse

Solvents are chemicals used to dissolve solids. The solvents used in paints, glues and lighter fluid are dangerous if they are breathed in. However, some people breathe them in for 'kicks' to give them a 'high' feeling. Using solvents in this way is known as **solvent abuse**. It can be addictive, like other forms of drug taking.

4 a) What is a solvent? **b)** What is solvent abuse?

People below the age of 16 are not allowed to buy substances containing these solvents. This law was made to try to reduce the number of people who die each year from breathing in solvent fumes.

5 Look at graph D.

   a) How many 15–19 years olds died from solvent abuse in 1986?
   b) How many 15–19 year olds died from solvent abuse in 1996?
   c) How do the numbers of people dying from solvent abuse in the 1990s differ from the numbers in the 1980s?
   d) Give one reason why there might be a difference between the numbers in the 1980s and 1990s.

Graph showing the number of deaths from solvent abuse each year.

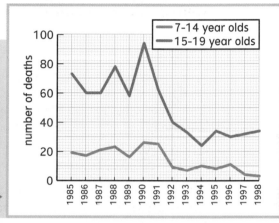

Solvents found in glues and paints are dangerous depressants which can stop the lungs and heart from working, causing severe brain damage. They also cause lung and liver damage. Sometimes the solvent can kill straight away. Solvents can also make people believe that they can fly and so some abusers throw themselves off high buildings and fall to their deaths.

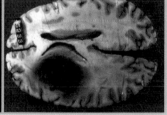

The brain on the left is normal. The brain on the right has been badly damaged by solvent abuse.

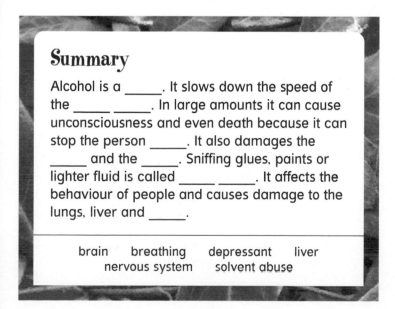

Summary

Alcohol is a _____. It slows down the speed of the _____ _____. In large amounts it can cause unconsciousness and even death because it can stop the person _____. It also damages the _____ and the _____. Sniffing glues, paints or lighter fluid is called _____ _____. It affects the behaviour of people and causes damage to the lungs, liver and _____.

brain	breathing	depressant liver
nervous system	solvent abuse	

6 'Short-term effects' describe what a drug does to someone's body straightaway.

   a) What are the short-term effects of drinking 10 units of alcohol?
   b) What do you think 'long-term effects' means?
   c) What are the long term effects suffered by an alcoholic?

7 People who abuse solvents often develop poor co-ordination. Why do you think this might be?

Further questions

1 Marimo duckweed, from Japan, is found as balls of stems and leaves. It is often found near the bottom of lakes. During the day, bubbles of gas collect inside the ball and make it float to the surface. The bubbles are released and it sinks again.

a) Name the gas that is found in the bubbles. (1)

b) What process produces this gas? (1)

c) i) Where in a lake would you expect to find balls of Marimo duckweed at night?

ii) Explain your answer. (2)

d) It has been found that increasing the light intensity (shining brighter light) makes more gas collect in the balls of weed. Look at the graph below.

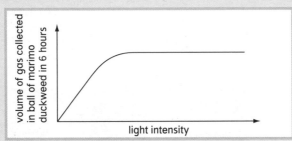

i) Increasing the light intensity does not keep on increasing the volume of collected gas. Explain why not. Use the words 'limiting factor' in your answer.

ii) Name one limiting factor in the experiment. (2)

2 Copy out this drawing of an animal cell.

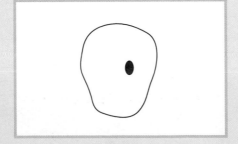

a) Add these labels to your drawing: cytoplasm, nucleus, cell membrane. (1)

b) Most of the chemical reactions in a cell happen in one part. Which part? (1)

c) Name one part that you would expect to find in a plant cell but not in an animal cell. (1)

3 Look at this drawing of a root hair cell. Its job is to absorb water.

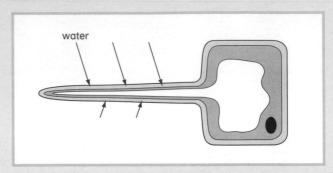

water

a) By which process does water get into this cell? Choose one word from these: (1)

> plasmolysis   evaporation   conduction
> osmosis   photosynthesis

b) How does the cell's shape help it do its job? (1)

c) Water is passed through other root cells until it reaches cells shaped like this:

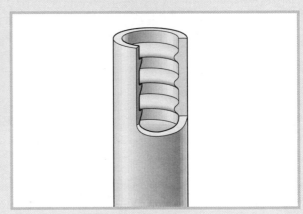

These cells take water up to the leaves.

i) What are these cells called?

ii) How are they adapted to their function?

iii) Which of these words is the correct name for transporting water up a plant? (3)

> diffusion   nutrition   transpiration
> transaction   nutrification

4 Look at this picture of dogs on a hot day. Dogs have the same organs that humans do.

This diagram shows the only two ways dogs can lose water

a) List three ways in which humans can lose water. (3)

b) Which one of these ways can dogs not lose water by? (1)

c) Name one waste product that needs to be removed from the body of a human and the body of a dog. (1)

d) Which organ gets rid of this waste? (1)

e) What is getting rid of waste like this called? Choose one word from these: (1)

| reproduction   excretion   respiration |
| secretion   incubation |

5 Copy and complete the sentences using words from the box. You may use the words once, more than once or not at all.

A _____ is a chemical that affects how your body works. There is a chemical like this in cigarette smoke called ____. Many people who smoke cannot stop because they are ____ on this chemical. The ___ in the smoke can cause ____ cancer and other lung diseases like _____ and _____. (7)

| bronchitis   colds   dependent   drug |
| emphysema   flu   lung   nicotine |
| **resting**   skin   tar   tartrazine |

6 Look at this diagram of the eye.

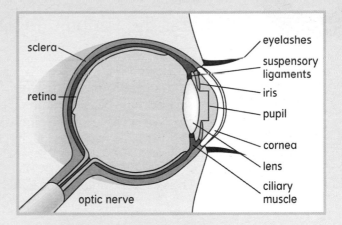

sclera, retina, optic nerve, eyelashes, suspensory ligaments, iris, pupil, cornea, lens, ciliary muscle

a) List two parts of the eye that must be transparent to let light through. (2)

b) The eye detects a stimulus. What stimulus does the eye detect? (1)

c) What organ does the optic nerve connect to? (1)

d) The organ you named in c) is part of an organ system.

  i) What is an organ system?

  ii) What organ system is the organ you named in part c) a part of? (2)

7 A plant that had been left in the dark for 24 hours was taken and put into a bell jar as shown. It was then left in bright light for 24 hours.

a) What gas is missing from the air inside the bell jar? (1)

b) Why was it important to seal the top of the bell jar? (1)

c) What substance, produced by photosynthesis, is starch made from? (1)

d) Why do the plants make starch? (1)

The Periodic Table

What is the Periodic Table?

There are 92 elements that are found naturally on Earth. 150 years ago not all these elements were known. Scientists were trying to put the elements that they did know about into a logical order.

 1 How many elements are found naturally on Earth?

In 1865, an English scientist called John Newlands arranged the elements in order of their atomic masses. The **atomic mass** is how much mass an atom of an element has compared with an atom of hydrogen. He found that every eighth element had similar properties.

Later, in 1869, a Russian scientist called Dimitri Mendeleyev continued Newlands' work and also put the elements in order of atomic mass. But this time he put them into a table. The elements were arranged in rows called **periods**. New rows were started so elements that were alike could line up in columns. These columns are called **groups**. The table was called the **Periodic Table**. Back in 1869 not all the elements were known and Mendeleyev left gaps in his table because he felt sure that there were some elements still to be discovered.

 2 What is a group?

3 What is a period?

Some elements were switched around to make them fit into the same group as other elements with similar properties. For example, argon has a bigger atomic mass than potassium so it would have come after potassium in the table. But it made more sense to switch them round so that potassium was in group 1 (the alkali metals) and argon was in group 0. This meant that some elements were no longer in order of atomic mass.

In the modern Periodic Table, the elements are arranged in order of **atomic number**. The atomic number is the number of **protons** in the atom. An atom of argon has 18 protons. This means the atomic number of argon is 18. An atom of potassium has an atomic number of 19. This means that an atom of potassium has 19 protons.

A John Newlands. **B** Dimitri Mendeleyev.

 **C** An atom has a nucleus the centre, which contains protons and neutrons. Tiny electrons orbit around the nucleus.

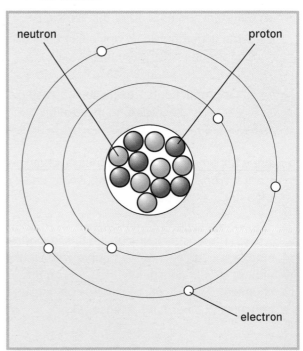

neutron

proton

electron

 4 Are elements arranged in order of atomic mass or atomic number?

Group numbers

Group	1	2												3	4	5	6	7	0 or 8
1st period													H hydrogen 1						He helium 2
2nd period	7 Li lithium 3	9 Be beryllium 4												11 B boron 5	12 C carbon 6	14 N nitrogen 7	16 O oxygen 8	19 F flourine 9	20 Ne neon 10
3rd period	23 Na sodium 11	24 Mg magnesium 12												27 Al aluminium 13	28 Si silicon 14	31 P phosphorus 15	32 S sulphur 16	35 Cl chlorine 17	40 Ar argon 18
4th period	39 K potassium 19	40 Ca calcium 20	45 Sc scandium 21	48 Ti titanium 22	51 V vanadium 23	52 Cr chromium 24	55 Mn manganese 25	56 Fe iron 26	59 Co cobalt 27	59 Ni nickel 28	64 Cu copper 29	65 Zn zinc 30		70 Ga gallium 31	73 Ge germanium 32	75 As arsenic 33	79 Se selenium 34	80 Br bromine 35	84 Kr krypton 36
5th period	85 Rb rubidium 37	88 Sr strontium 38	89 Y yttrium 39	91 Zr zirconium 40	93 Nb niobium 41	96 Mo molybdenum 42	98 Tc technetium 43	101 Ru ruthenium 44	103 Rh rhodium 45	106 Pd palladium 46	108 Ag silver 47	112 Cd cadmium 48		115 In indium 49	119 Sn tin 50	122 Sb antimony 51	128 Te tellurium 52	127 I iodine 53	131 Xe xenon 54
6th period	133 Cs caesium 55	137 Ba barium 56	139 La lanthanum 57	178 Hf hafnium 72	181 Ta tantalum 73	184 W tungsten 74	186 Re rhenium 75	190 Os osmium 76	192 Ir iridium 77	195 Pt platinum 78	197 Au gold 79	201 Hg mercury 80		204 Tl thallium 81	207 Pb lead 82	209 Bi bismuth 83	209 Po polonium 84	210 At astatine 85	222 Rn radon 86

the alkali metals   the transition metals

D The modern Periodic Table. The elements are arranged in order of atomic number.

E Some of these are metals and some are non-metals. Can you tell which is which?

Of the 92 elements found naturally on Earth, more than three-quarters are **metals** and the rest are **non-metals**.

We can draw a line on the Periodic Table to divide the metals and non-metals. The metals are on the left-hand side of the line. The non-metals are on the right hand side of the line.

Summary

In the Periodic Table, elements are arranged in order of _____ number. Each row in the table is called a _____. Elements with the same properties fit into columns called _____. More than three-quarters of the elements are _____. These can be found in group 1 (the _____ metals), group 2 and a block in the middle called the _____ metals. The rest of the elements are _____-_____.

alkali   atomic   groups   metals
non-metals   period   transition

? 5 What fraction of the elements are metals?

6 a) Write down the names of the elements in the 2nd period.

b) Write down the names of the elements in group 1.

c) Write down the names of the elements in the 3rd period.

7 Write down the names of 5 metals with similar properties.

Metals and non-metals

What are the differences between metals and non-metals?

When you think of a metal or a non-metal there will be words you use to describe it. These words are its **properties**.

1 a) What words could be used to describe a metal like iron?

b) Carbon is a non-metal used in pencil leads. What words could be used to describe a non-metal like carbon?

Physical properties of non metals

Most non-metals are:

- dull
- poor conductors of electricity
- poor conductors of heat.

Physical properties of metals

All metals are:

- shiny
- good conductors of heat
- good conductors of electricity.

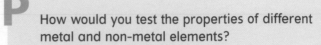

P How would you test the properties of different metal and non-metal elements?

- Are they shiny or dull?
- Do they conduct electricity or not?
- Do they conduct heat well or not?

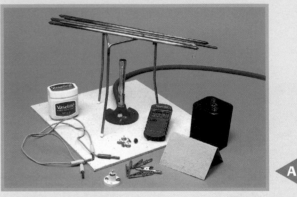

 A

All metals have these properties. Diagram B shows how the physical properties of group 1 metals are different from transition metals.

 B

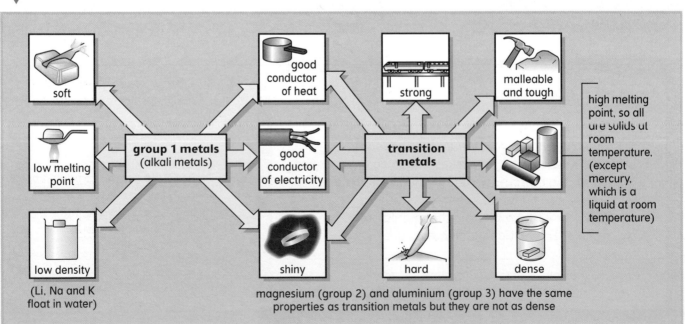

(Li, Na and K float in water)

magnesium (group 2) and aluminium (group 3) have the same properties as transition metals but they are not as dense

! Gold is so malleable that it can be beaten out into a film less than 0.009 mm thick.

Chemical properties of metals

Group 1 metals:

- are very reactive
- react very quickly with oxygen
- react quickly with water releasing hydrogen
- form hydroxides, which dissolve in water to form alkaline solutions
- react with non-metals to form ionic compounds that are white solids. These white solids dissolve in water to form colourless solutions.

This copper roof has reacted slowly over time with air (oxygen) and water. The compound formed is green.

Transition metals:

- are much less reactive than group 1 metals
- react very slowly with oxygen and water
- form coloured compounds (these can be seen in the different coloured glazes in pottery and on old copper roofs).

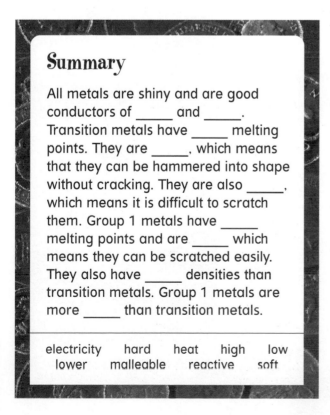

Summary

All metals are shiny and are good conductors of _____ and _____. Transition metals have _____ melting points. They are _____, which means that they can be hammered into shape without cracking. They are also _____, which means it is difficult to scratch them. Group 1 metals have _____ melting points and are _____ which means they can be scratched easily. They also have _____ densities than transition metals. Group 1 metals are more _____ than transition metals.

electricity	hard	heat	high	low
lower	malleable	reactive	soft	

3 a) Write down two properties that both group 1 metals and transition metals have.

  b) Write down two differences between group 1 metals and transition metals.

4 When group 1 metals are freshly cut, they are shiny inside. But they quickly tarnish and turn dull again. Explain why this happens.

Uses of metals

Why are metals so useful?

A

Metals have lots of uses. You use metals every day of your life. Think about having breakfast. Your cereal spoon is probably made of metal, so are the radiators, the taps in the sink, parts of the microwave, and the fridge. There are also many other things made from metals like grill pans and satellite dishes. All of these metals before you even leave your house!

Transition metals in particular are very useful.

1 Write down 5 things in your house that are made from metals.

2 Write down 3 things outside your house that are made from metals.

Copper

Wires and cables are made of copper with a plastic coating on them. Copper is used to make electrical wiring because it can easily be drawn into wires and is an excellent conductor of electricity.

B

C The Statue of Liberty in New York Harbour, USA, is made out of copper. It is hollow inside.

Iron

Iron is used to make gates because it is malleable. It can easily be hammered into shape. It is also very strong. When it is mixed with carbon to make steel it is even stronger. Steel can be used to make cars and the girders that are used in buildings and bridges.

Transition metals are useful as **catalysts** in industry. Catalysts speed up chemical reactions. Iron is used as a catalyst in the reaction to make a gas called ammonia. Ammonia is then used to make fertilisers. Platinum is used in catalytic converters that are fitted to car exhausts. It cuts down the amount of pollution in car exhaust fumes.

This aeroplane is made of an alloy of aluminium, copper and magnesium **F**

3 What properties make copper useful for electrical wiring?

4 What properties make iron useful for cars?

Summary

Metals can be used to make many things. Copper is used to make _____ because it is a good conductor of electricity. Iron is used to make _____ because it is strong. Transition metals can be used as _____ to speed up chemical reactions. An _____ is a mixture of metals.

alloy   cars   catalysts   wires

Alloys

Aluminium is light and can be used to make aeroplanes. But aeroplanes are not made of pure aluminium because it is not strong enough. The wings might bend too much and snap off! Mixing aluminium with other metals like copper and magnesium makes it harder, stronger and stiffer. A mixture of metals is called an **alloy**.

5 What is an alloy?

6 **a)** Name 2 things made from iron.

  b) What property of iron makes it useful for making gates?

7 Why are alloys used to make aeroplanes, instead of just using the separate metals?

8 What properties do you think aluminium has that makes it useful for overhead power lines?

Metals and air

What is made when metals react with air?

A *Magnesium burning.*

B *Freshly cut sodium.*

Different elements can join together to make a **compound**. When magnesium burns, it joins up with oxygen from the air to make a new compound. When metals join up with oxygen they form **oxides**. The new compound made when magnesium joins up with oxygen is called magnesium oxide.

The **word equation** is:

magnesium + oxygen ⟶ magnesium oxide

These are called **reactants**. They react together to make a new chemical.

The new chemical made is called the **product**.

> **?** 1 Which gas from the air do metals join up with when they are heated or burned?
>
> 2 When magnesium reacts with oxygen, what compound is made?

Reactants and products in an equation can be written in a short-hand way. You can re-write the word equation above using the **chemical formula** for each substance. So now you can write a **symbol equation** for burning magnesium:

$$2Mg + O_2 \longrightarrow 2MgO$$

2 atoms of magnesium.

Oxygen in the air exists as molecules. Each molecule is 2 oxygen atoms joined together.

There is 1 oxygen atom for each magnesium atom.

When writing symbol equations, you may need to put numbers on the front of a formula. This makes sure that you have the same numbers of atoms on each side of the equation.

Group 1 metals

If you cut lithium, sodium or potassium with a knife (yes, they are soft!) they are shiny inside. But if you look closely, they start turning dull very quickly. This is because the metal is reacting with oxygen to form an oxide. The dull colour is the oxide.

The word and symbol equations are:

lithium + oxygen $\longrightarrow$ lithium oxide

$4Li + O_2 \longrightarrow 2Li_2O$

sodium + oxygen $\longrightarrow$ sodium oxide

$4Na + O_2 \longrightarrow 2Na_2O$

Potassium reacts more quickly than sodium. This means that potassium is more **reactive** than sodium. Sodium reacts more quickly than lithium. This means that sodium is more reactive than lithium.

3 Why does sodium change colour after it has been freshly cut?

4 Look at the word equations above. Now write a word equation for potassium reacting with oxygen.

Transition metals like zinc, iron and copper need to be heated to make them react with the oxygen in the air. Group 1 metals are more reactive and do not need to be heated.

5 Which metal is more reactive, potassium or copper?

Summary

When different elements join together they make _____. Metals react with _____ in the air to form_____ . Some metals are more _____ than others. Lithium, sodium and potassium react very _____ with oxygen. Other metals like zinc and _____ need to be _____ to make them react with oxygen. By seeing how quickly metals react with oxygen, we can put them in order of _____.

compounds	heated	iron	quickly
oxides	oxygen	reactive	reactivity

P How would you find out how quickly other metals react with oxygen?

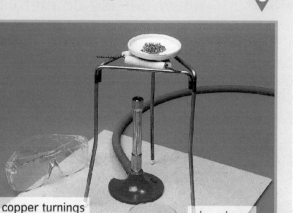

copper turnings
zinc pieces
magnesium ribbon

Metals can be put in a list, in order of **reactivity**. The most reactive metal is at the top of the list. The least reactive metal is at the bottom.

potassium

sodium

lithium

zinc

iron

copper

6 What is made when potassium reacts with oxygen from the air?

7 a) What is made when zinc reacts with oxygen from the air?
  b) Write a word equation for this reaction.

8 a) What is made when copper reacts with oxygen from the air?
  b) Write a word equation for this reaction.

9 Write symbol equations for the two reactions in questions 7 and 8.

Metals and water

How quickly do metals react with water?

Some metals react quickly with water. Others react slowly and some, like gold, do not react at all. You have probably seen what happens to a metal when it has been in contact with air and water for a long time. It **corrodes**. When iron corrodes, we call it **rusting**.

 1 What happens to metals when they react slowly with air and water?

When reactive metals like sodium react with water, a **metal hydroxide** is made.

The word equation is:

metal + water ⟶ metal hydroxide + hydrogen
e.g. sodium + water ⟶ sodium hydroxide + hydrogen

The symbol equation is:

$$2Na + 2H_2O \longrightarrow 2NaOH + H_2$$

The sodium moves about on the surface of the water very quickly. There is lots of fizzing because a gas is being made. The gas is **hydrogen**.

Potassium reacts in a similar way to sodium but this reaction is more **violent**. Potassium is more reactive than sodium and sodium is more reactive than lithium.

B Sodium (left) and potassium (right) reacting with water

A

2 How can you tell that potassium is more reactive than sodium?

3 What gas is made when a metal reacts quickly with water?

4 What else is made when a metal reacts quickly with water?

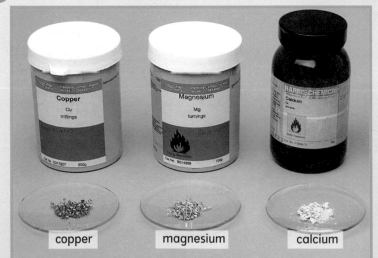

copper   magnesium   calcium

C How would you find out which metal reacts quickest with water: calcium, magnesium or copper?

Calcium reacts quickly with water, but not violently.

$$\text{calcium} + \text{water} \longrightarrow \text{calcium hydroxide} + \text{hydrogen}$$
$$\text{Ca} + 2H_2O \longrightarrow Ca(OH)_2 + H_2$$

Magnesium reacts very slowly with cold water. We can speed up the reaction by using hot water.

$$\text{magnesium} + \text{water} \longrightarrow \text{magnesium hydroxide} + \text{hydrogen}$$
$$\text{Mg} + 2H_2O \longrightarrow Mg(OH)_2 + H_2$$

The reaction works best if we heat the magnesium with steam instead. When a metal reacts with steam instead of water, a **metal oxide** is made instead of a metal hydroxide. Hydrogen gas is still made. The magnesium reacts strongly with steam leaving a white powder which is magnesium oxide.

$$\text{magnesium} + \text{steam} \longrightarrow \text{magnesium oxide} + \text{hydrogen}$$
$$\text{Mg} + H_2O \longrightarrow MgO + H_2$$

Iron also reacts with steam to form iron oxide and hydrogen. Copper, silver and gold do not react with water or steam.

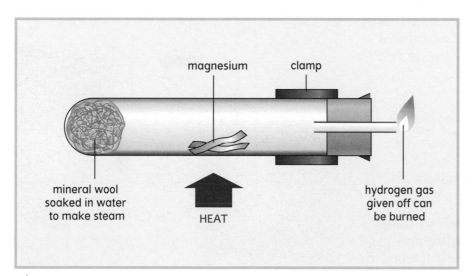

 D

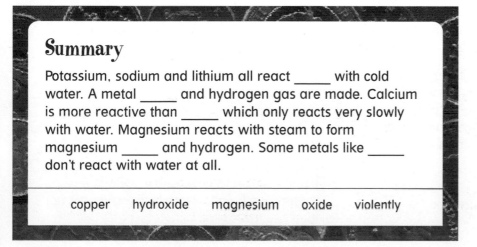

Summary

Potassium, sodium and lithium all react _____ with cold water. A metal _____ and hydrogen gas are made. Calcium is more reactive than _____ which only reacts very slowly with water. Magnesium reacts with steam to form magnesium _____ and hydrogen. Some metals like _____ don't react with water at all.

copper   hydroxide   magnesium   oxide   violently

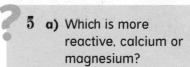

5 a) Which is more reactive, calcium or magnesium?
  b) Explain your answer.

6 What are the products when magnesium reacts with water?

7 What are the products when magnesium reacts with steam?

We can put the metals we have met so far in order of reactivity, as follows:

potassium

sodium

lithium

calcium

magnesium

iron

copper

silver

gold

8 Which is more reactive, lithium or calcium?

9 What gas is made when calcium reacts with water?

10 a) What gas is made when lithium reacts with water?
   b) What else is made when lithium reacts with water?

C6 Metals and acids

What happens when metals react with acids?

Metals that are less reactive than calcium react very slowly with water and some of them do not react at all. However, metals react more quickly with dilute acid. This makes it easier to see which metals are the most reactive.

When a metal reacts with an acid a **salt** is made. Hydrogen gas is also made. For example, if you put a small piece of magnesium into dilute hydrochloric acid you can see bubbles of gas. The bubbles are hydrogen gas.

dilute hydro-chloric acid

magnesium

bubbles of hydrogen gas

The word equations are

magnesium + hydrochloric acid $\longrightarrow$ magnesium chloride + hydrogen

$$Mg + 2HCl \longrightarrow MgCl_2 + H_2$$

A

B

The salt made in this reaction is called magnesium chloride. Another salt you may have heard of is sodium chloride. This is the salt you put on your food and is often called common salt.

1 What two chemicals are made when a metal is added to dilute acid?

2 What is the name of the salt made when magnesium is added to dilute hydrochloric acid?

The hydrogen gas can be collected easily by using a test tube as shown in diagram B. Hydrogen is lighter than air, so it rises upwards and is trapped in the empty test tube. You can't see or smell hydrogen gas but you can do a test to see if it is there.

If a test tube of hydrogen is held next to a lighted splint, the hydrogen will burn in the air. It burns very quickly and makes a squeaky 'pop'.

You can compare the reactivity of different metals by seeing how quickly hydrogen is made when they react with acid. The faster the hydrogen is made, the more reactive the metal. Some results from the experiment are shown in diagram C.

3 Why is it easy to collect hydrogen in an upside down test tube?

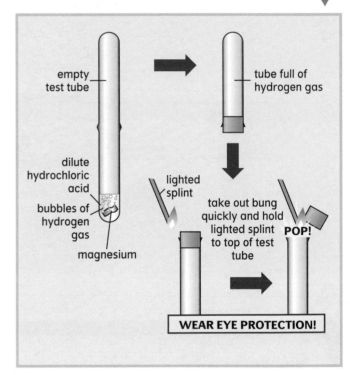

empty test tube

tube full of hydrogen gas

dilute hydrochloric acid

bubbles of hydrogen gas

magnesium

lighted splint

take out bung quickly and hold lighted splint to top of test tube

POP!

WEAR EYE PROTECTION!

! Airships and balloons used to be filled with hydrogen because it is lighter than air. Unfortunately, hydrogen is also very flammable and airships were in danger of exploding into flames. Helium is now used instead. It is also lighter than air but is unreactive and does not burn like hydrogen.

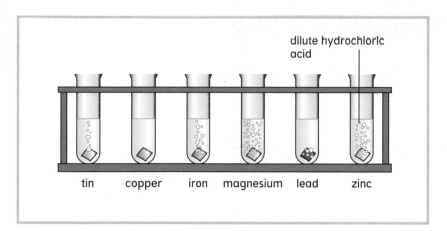
dilute hydrochloric acid

tin   copper   iron   magnesium   lead   zinc

P How would you investigate which metal reacts the fastest with acid?

D

The acid in the test tubes containing tin and lead has to be warmed before these two metals will react. Copper does not react with dilute acid, even if it is warmed. **C**

The other metals do react. For example:

zinc + hydrochloric acid $\longrightarrow$ zinc chloride + hydrogen

Zn + 2HCl $\longrightarrow$ $ZnCl_2$ + H_2

Very reactive metals like potassium explode in dilute acid.

Summary

When a metal is added to dilute acid a _____ and hydrogen gas are made. If zinc is added to dilute hydrochloric acid, the salt is called _____ _____ . To test for hydrogen, a lighted splint is used. The hydrogen burns with a squeaky _____. Some metals are more _____ than others. For example, _____ is more reactive than zinc.

magnesium   pop   reactive
salt   zinc chloride

? 4 Which test tubes have to be warmed to make the metal react with the acid?

5 a) Which metal is most reactive?
b) How can you tell?

6 a) Which metal is least reactive?
b) How can you tell?

From the results we can put the metals in order of reactivity:

magnesium   most reactive
zinc
iron
tin
lead
copper   least reactive

? 7 What is the test for hydrogen?

8 a) Which is more reactive, zinc or iron?
b) How can you tell?

9 What is the name of the salt made when iron is added to dilute hydrochloric acid?

10 Copy and complete the word equation:

iron + hydrochloric acid $\rightarrow$ _____ _____ + _____

C7 The Reactivity Series

What is the Reactivity Series and what does it mean?

We have seen that some metals are more reactive than others. If you look back to see how reactive the metals were with oxygen from the air, water and acid you can put all the results together. From these results you can put the metals into a list called the **Reactivity Series**. The Reactivity Series is like a league table for metals. The most reactive metal is at the top of the league table and the least reactive metal is at the bottom.

1 What is the Reactivity Series?

2 Look at table A.

    a) Which metal is most reactive?
    b) Which metal is least reactive?
    c) Which is more reactive, calcium or iron?

We can put all the results from the last three topics together:

Metal	Symbol	Reaction when heated in air (oxygen)	Reaction with cold water	Reaction with dilute acid
potassium	K		Fizzes violently, giving off hydrogen; forms metal hydroxide solutions.	Very violent reaction (explodes — extremely dangerous).
sodium	Na			
lithium	Li			
calcium	Ca	Burn brightly forming an oxide.	Very slow reaction.	Fizz giving off hydrogen gas; form metal salts.
magnesium	Mg			
aluminium	Al		No reaction except for slow rusting of iron; all react with steam.	
zinc	Zn			
iron	Fe			
tin	Sn	Oxide layer forms but metal does not burn.	No reaction, even with steam.	React very slowly giving off hydrogen.
lead	Pb			
copper	Cu			No reaction.
silver	Ag			
gold	Au	No reaction.		
platinum	Pt			

most reactive

metal	symbol
potassium	K
sodium	Na
lithium	Li
calcium	Ca
magnesium	Mg
aluminium	Al
zinc	Zn
iron	Fe
tin	Sn
lead	Pb
copper	Cu
silver	Ag
gold	Au
platinum	Pt

least reactive

A The Reactivity Series.

3 Which metals form oxides without burning when heated in air?

4 Which metals react with cold water giving off hydrogen gas?

5 Which metals don't react with water or steam?

We can use the Reactivity Series to make predictions about chemical reactions. Look at diagram C.

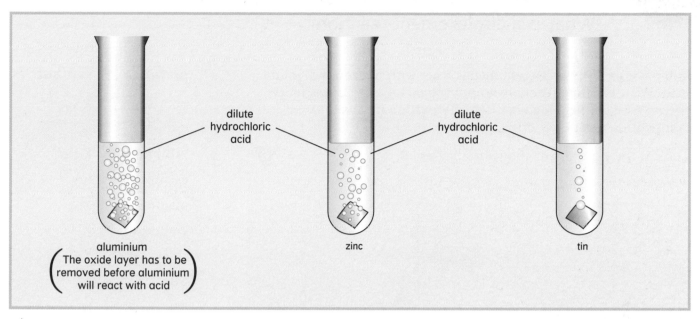

Aluminium is higher in the Reactivity Series than zinc, so it is more reactive. If you put pieces of each metal in acid they will react and give off hydrogen gas. We can show that aluminium is more reactive by drawing more bubbles. Tin is lower than aluminium and zinc in the Reactivity Series. We can predict that there will be fewer bubbles than with aluminium or zinc. We can show this by drawing fewer bubbles on the diagram than there are for zinc.

Summary

The _____ _____ is a list of metals in order of reactivity. It is like a _____ table for metals. The _____ reactive metal is placed at the top of the list. The least reactive metal is placed at the _____ of the list. Very reactive metals like sodium react very quickly with air and water and _____ if put into acid. Metals halfway down the series like zinc and _____ don't react with water but do react with _____. They fizz in acid giving off _____ gas. Metals near the bottom of the series like silver and _____ do not react with air, water or even acid.

bottom	explode	gold
hydrogen	iron	league
most	Reactivity Series	steam

6 Draw a test tube to show iron reacting with dilute acid.

7 a) Which metals don't react with water but do react with steam?
   b) What is made when these metals react with steam?

8 a) Which metals fizz when they react with acid giving off hydrogen gas?
   b) What else is made when these metals react with acid?

9 Copy and complete this word equation:

zinc + hydrochloric acid ⟶

_____ + _____

10 Write a symbol equation for the reaction in question 9.

Displacement reactions

What is a displacement reaction?

You have seen how different metals react with oxygen, water and acid. We used these reactions to put the metals into a Reactivity Series. We can also judge reactivity by putting metals into competition with each other.

Look at the experiment shown in diagram B:

B

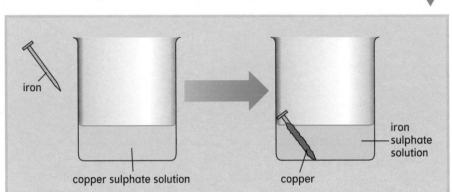

| iron | iron sulphate solution |

copper sulphate solution — copper

The word and symbol equations are:

iron + copper sulphate ⟶ iron sulphate + copper

Fe + CuSO₄ ⟶ FeSO₄ + Cu

(grey)   (blue)   ⟶   (green)   (red-brown)

Iron is higher in the Reactivity Series and 'kicks out' or **displaces** copper from its solution. You can see the red-brown copper coating the nail. The iron has taken its place in solution and forms iron sulphate solution, which is green. A more reactive metal can displace ('kick out') a less reactive metal from its solution. This type of reaction is called a **displacement reaction**.

P How would you investigate whether other metals carry out displacement reactions?

C

| zinc sulphate | copper sulphate |
| iron sulphate | magnesium sulphate |

most reactive

metal	symbol
potassium	K
sodium	Na
lithium	Li
calcium	Ca
magnesium	Mg
aluminium	Al
zinc	Zn
iron	Fe
tin	Sn
lead	Pb
copper	Cu
silver	Ag
gold	Au
platinum	Pt

least reactive

A *The Reactivity Series*

?1 What is a displacement reaction?

2 Why does iron displace copper from solution?

If we put copper into iron sulphate solution, there would be no reaction, because copper is below iron in the Reactivity Series. Copper cannot displace iron. This means it cannot kick iron out of its solution and take its place.

?3 Why can't copper displace iron from its solution?

Iron versus copper

We can put metals into competition with each other to win oxygen. If we heat iron metal and copper oxide together, the iron and the copper will compete for the oxygen. Copper has the oxygen to start with. Iron is higher in the Reactivity Series so it takes the oxygen from copper.

If you mix a spatula of iron filings and copper oxide in a test tube and heat it strongly, you will see a red glow after a while. This tells you there is a reaction happening.

After a few minutes, if you pour the mixture into an evaporating basin, you can see small pieces of red–brown copper.

The iron has displaced the copper and has won the competition for oxygen. It takes the oxygen from copper because it is higher in the Reactivity Series. It is more reactive.

$$\text{iron} + \text{copper oxide} \longrightarrow \text{iron oxide} + \text{copper}$$

If we heat copper with iron oxide there would be no reaction. The iron will keep the oxygen because it is more reactive than copper.

$$\text{copper} + \text{iron oxide} \longrightarrow \text{no reaction}$$

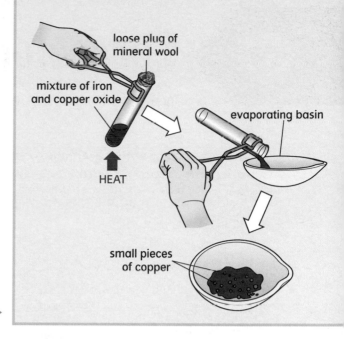

loose plug of mineral wool

mixture of iron and copper oxide

HEAT

evaporating basin

small pieces of copper

D

E

?

4 a) Which metal wins the competition, iron or copper?

b) Explain why this happens.

5 How could you tell that there was a reaction happening?

Summary

A more reactive metal can _____ a less reactive metal from its solution. For example, _____ will displace iron in a reaction. But if iron were added to magnesium sulphate solution, it would not be able to displace _____ from its solution. If iron and copper oxide are mixed together and heated, _____ will win the competition for oxygen. This is because iron is _____ than copper in the Reactivity Series. Copper will be _____ by iron.

displace	displaced	higher
iron	magnesium	zinc

?

6 Look at table A. Predict whether there will be a reaction between these pairs. Write down 'yes' or 'no' for each of your answers.

a) Magnesium and zinc sulphate.

b) Copper and zinc oxide.

c) Aluminium and iron sulphate.

d) Explain each of your answers for a), b) and c).

e) Write word equations for any reactions that work.

7 Write symbol equations for the reactions in question 7 that work.

Corrosion of metals

What is corrosion and how can we prevent it?

When metals react with water and oxygen from the air they can turn into oxides or hydroxides. This is called **corrosion**.

Corrosion is not useful. If a metal bridge corrodes it becomes weak and might fall down.

1 What is corrosion?
2 What two things make metals corrode?

Iron (or steel) corrodes much faster than other transition metals. Corroded metals have to be replaced. This costs the country millions of pounds each year.

 B

Preventing the corrosion of iron (or steel)

When iron corrodes, it is called **rusting**. Oxygen and water are both needed to make iron rust. If oxygen and water can be kept away from iron or steel then it won't rust.

3 What is rusting?

Coating the metal with paint, oil (or grease), or plastic forms a barrier and stops oxygen and water getting to the iron or steel.

Iron can also be coated with a more reactive metal like zinc. Zinc is more reactive than iron. Oxygen and water will react with the zinc instead of the iron. The zinc sacrifices itself for the iron. This is called **sacrificial protection**.

Magnesium is sometimes used instead of zinc.

A

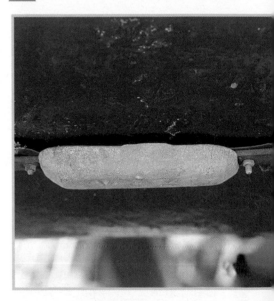
C These magnesium bars protect the ship's hull from rust. They corrode instead of the steel.

4 a) What metals are used for sacrificial protection?
b) Explain why they can be used.

Chromium and nickel can be added to steel to make an alloy called **stainless steel**. This does not rust easily. It is useful for small items like knives and forks, but is expensive.

5 Write down three ways that rusting can be stopped.

6 What is the advantage of stainless steel over ordinary steel?

Aluminium is quite a reactive metal, but doesn't corrode very quickly. This is because a thin layer of aluminium oxide forms on the surface. This oxide layer stops oxygen and water getting to the metal underneath.

7 Why is stainless steel not used to make cars?

8 a) Aluminium cars do not corrode. Why not?

b) Why is pure aluminium not used to make cars? (Hint: you may need to look back at topic C3.)

 These knives and forks are made from stainless steel.

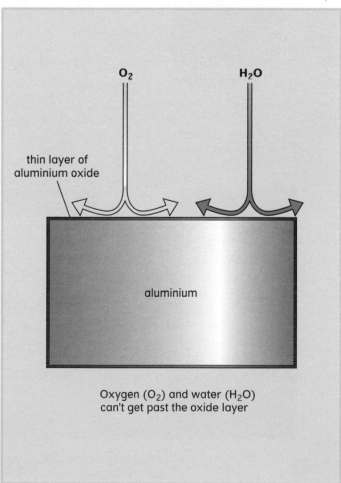

thin layer of aluminium oxide

O₂   H₂O

aluminium

Oxygen (O₂) and water (H₂O) can't get past the oxide layer

Summary

When metals corrode they react with _____ and _____ from the air. When iron or steel corrode, it is called _____ We can prevent this by using a barrier method; covering the metal with a layer of _____, _____ or plastic. Iron or steel can also be coated with a more reactive metal to stop them rusting. This is called _____ _____. Chromium and _____ are added to steel to make an alloy called _____ _____. Aluminium does not corrode because it has a thin layer of aluminium _____ on the surface. This stops oxygen and water getting to the metal underneath.

grease	nickel	oxygen	oxide
paint	rusting	sacrificial protection	
stainless steel	water		

C10 Extraction of metals

Where do we find metals?

The Earth's crust is a thin layer of rock covering the whole of the planet. We call the Earth's crust the ground. Metals are found in the Earth's crust.

 1 Where do we find metals?

Metals at the bottom of the Reactivity Series are unreactive. They can be found as the metals themselves, not joined up with other elements as compounds. Metals like gold and platinum are found on their own. We say that these metals are found in their **native** state. Copper is sometimes found in its native state.

 2 Gold is found in its native state. What does this mean?

3 Why is gold found in its native state?

 C Copper   **D** Gold

We don't find many metals just lying around on the ground. Most metal elements are joined up with other elements to make compounds. The metal compounds are mixed with rock and this mixture is called an **ore**. An ore contains enough of the metal compound to make it worth extracting. An ore often contains a metal oxide. To extract the metal, we have to remove oxygen from the metal oxide.

A

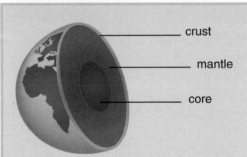

crust

mantle

core

B

 4 What is an ore?

Panning for gold: a metal found in its native state. **E**

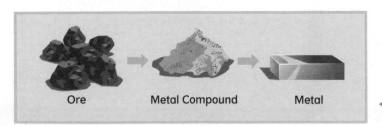

Ore   Metal Compound   Metal

F

104

Extraction of metals and the Reactivity Series

To extract a metal from its ore, the oxygen must be taken away from the metal oxide. Taking oxygen away from something is called **reduction**.

There is a link between a metal's place in the Reactivity Series and how easy it is to extract from its ore. The more reactive a metal is, the more difficult it is to extract from its ore. Reactive metals like aluminium and sodium want to stay joined up with other elements in their ores. This makes it difficult to extract very reactive metals.

We use electricity in a process called **electrolysis** to extract very reactive metals, like aluminium, from their ores. Less reactive metals, like iron, can be extracted by heating the metal oxide with carbon. Carbon is more reactive than iron, so it 'grabs' the oxygen and displaces the iron (a displacement reaction).

e.g.   iron oxide + carbon $\longrightarrow$ iron + carbon dioxide

$$2FeO + C \longrightarrow 2Fe + CO_2$$

Electrolysis can be used for metals lower down in the Reactivity Series, but it is not used because it is a very expensive process.

potassium		**hard to extract**
sodium		
lithium	Use electrolysis.	
calcium		
magnesium		
aluminium		
(carbon)		
zinc		**getting harder to extract**
iron	Ores are heated with carbon.	
tin		
lead		
(hydrogen)		
copper		
silver	Found in native state (although copper is sometimes found as an ore).	
gold		
platinum		**easy to extract**

 Carbon and hydrogen are non-metals but they can be added to the Reactivity Series.

?

5 What is reduction?

6 Which metal is easier to extract from its ore, iron or aluminium?

! There is more aluminium in the Earth's crust than any other metal. Unfortunately it is very difficult to get aluminium from its ore.

?

7 a) Which is more difficult to extract, sodium or lead?
  b) Explain your answer.

8 Metals can be extracted in two ways. Heating the oxide with carbon is one way. What is the other way?

9 a) Which method would you use to extract tin from its ore?
  b) Which method would you use to extract sodium from its ore?
  c) Explain your choices for a) and b).

Summary

Metals are found in rocks called _____ which usually contain the metal or a metal _____. Metals at the bottom of the Reactivity Series like gold and _____ are found in their _____ state. This means that they are found as the metals themselves, and not joined up with other elements because they are _____. The higher up the Reactivity Series a metal is, the more _____ it is to extract.

difficult   native   ores   oxide
platinum   unreactive

Extracting metals with carbon

How can carbon be used to extract metals from their ores?

Carbon is a non-metal but we can put it in the Reactivity Series. Carbon is less reactive than aluminium but more reactive than zinc. That means it slots in between those two metals as shown in diagram A.

Carbon can **displace** any metal below aluminium in the Reactivity Series. This means we can use carbon to extract metals below aluminium from their ores. For example, iron is less reactive than carbon, so carbon can be used to extract iron from its ore. Carbon 'grabs' the oxygen from the ore and becomes carbon dioxide.

Carbon is found in the form of coal so it is cheap. Remember, the ore usually contains the metal oxide, so all we have to do is heat the metal oxide with carbon to get the metal. The reaction is a **displacement reaction**.

1 Where is carbon in the Reactivity Series?

2 **a)** Name a metal that can be extracted by using carbon.

b) Why is this method less expensive than electrolysis?

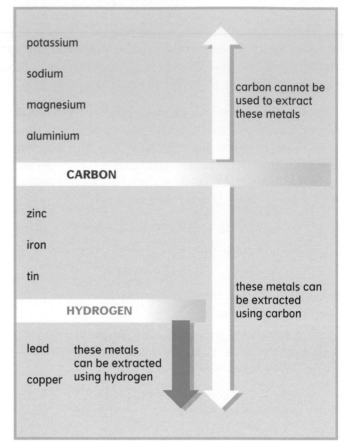

potassium

sodium

magnesium

aluminium

CARBON

zinc

iron

tin

HYDROGEN

lead

copper

carbon cannot be used to extract these metals

these metals can be extracted using carbon

these metals can be extracted using hydrogen

A

B

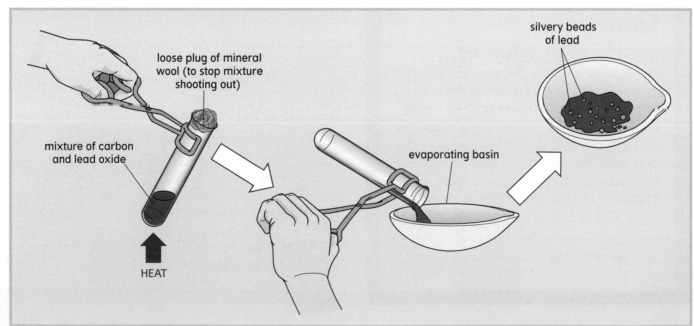

silvery beads of lead

loose plug of mineral wool (to stop mixture shooting out)

mixture of carbon and lead oxide

evaporating basin

HEAT

Hydrogen is also a non-metal and can be put into the Reactivity Series like carbon. It is more reactive than lead and could be used to extract lead from its ore. However, it is much lower down in the reactivity series than carbon so is not as useful. Carbon is easier to use than hydrogen as it is a solid. Carbon is usually used to extract metals from their ores instead of hydrogen.

Extracting lead from its ore

Lead oxide is a yellow powder. We can get it from an ore containing lead.

Carbon is more reactive than lead. If we heat lead oxide with carbon, the carbon takes oxygen away from lead oxide in a displacement reaction. Carbon has displaced lead from lead oxide. Taking oxygen away from a compound is called **reduction**.

Carbon has **reduced** lead oxide to lead.

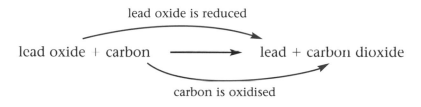

lead oxide is reduced

lead oxide + carbon ⟶ lead + carbon dioxide

carbon is oxidised

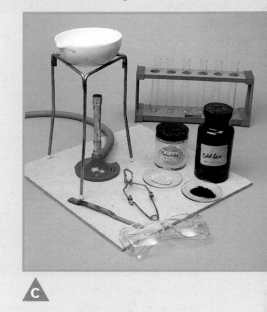

P How can you extract copper from its ore using carbon?

C

When oxygen is added to a chemical we say it has been **oxidised**. Adding oxygen to something is called **oxidation**. Carbon is oxidised when it takes oxygen away from lead oxide.

Summary

Carbon is a non-metal but it can be put into the _____ Series. It slots in between aluminium and _____. We can use carbon to extract metals from their _____. Carbon can be used to _____ any metal below aluminium in the Reactivity Series. Iron, which is less _____ than carbon can be extracted from its ore by heating it with carbon. Taking oxygen away from a metal oxide is called _____. The oxygen is added to the carbon; this is called _____.

extract   ores   oxidation
reactive   Reactivity
reduction   zinc

? 3 Why is it possible to use carbon to extract lead from its ore?

4 What is reduction?

5 What is oxidation?

6 a) Which metals can be extracted from their ores by using carbon?
   b) Why can carbon displace these metals?

7 a) Can carbon be used to extract aluminium from its ore?
   b) Explain your answer.

8 Look at this word equation

   zinc oxide + carbon ⟶ zinc + carbon dioxide

   a) What has been oxidised?
   b) What has been reduced?
   c) Write a symbol equation for the reaction.

C12 Getting iron from its ore

How do we get iron from its ore?

 From this …

… to this. B

Iron is an extremely useful metal and we use a lot of it. To get iron we need three raw materials:

1. **Iron ore** — mainly haematite which contains iron oxide

2. **Coke** — a cheap form of carbon made from coal

3. **Limestone** — to get rid of impurities.

The three raw materials are crushed and heated together in a **blast furnace**. It is called a blast furnace because hot air is blasted in at the bottom. It is very hot inside a blast furnace; the temperature can be up to 1900 °C.

 1 What are the raw materials used for extracting iron from its ore?

Reactions in the blast furnace

The coke (carbon) burns. It reacts with oxygen from the hot air blasted in and makes carbon dioxide. This reaction produces lots of heat and makes the blast furnace even hotter.

carbon + oxygen $\longrightarrow$ carbon dioxide

$$C + O_2 \longrightarrow CO_2$$

The carbon dioxide formed reacts with hot coke (carbon) to make **carbon monoxide** gas.

carbon dioxide + carbon $\longrightarrow$ carbon monoxide

$$CO_2 + C \longrightarrow 2CO$$

C A blast furnace.

The carbon monoxide then **reduces** the iron oxide to iron. Carbon monoxide is called a **reducing agent**.

iron oxide + carbon monoxide $\longrightarrow$ iron + carbon dioxide

$$Fe_2O_3 + 3CO \longrightarrow 2Fe + 3CO_2$$

The carbon monoxide is itself oxidised when it joins up with oxygen from the iron ore.

Limestone reacts with the impurities in the iron ore to make a liquid waste called **slag**.

2 What substance reduces the iron ore to iron?

3 What is a reducing agent?

As the temperature in the furnace is very high, the iron formed is a liquid and sinks to the bottom of the furnace. The slag is also a liquid, but it is lighter (less dense) than liquid iron and so floats on top of it. The slag and molten iron are run off separately. When the slag has cooled and solidified it is used for making roads and making breeze blocks for buildings.

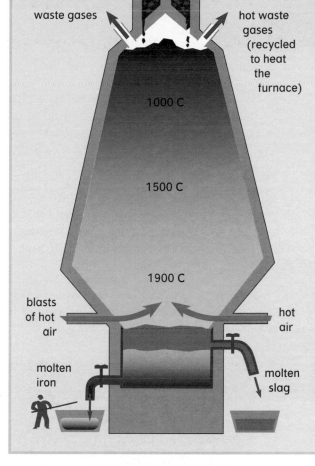

The blast furnace. **D**

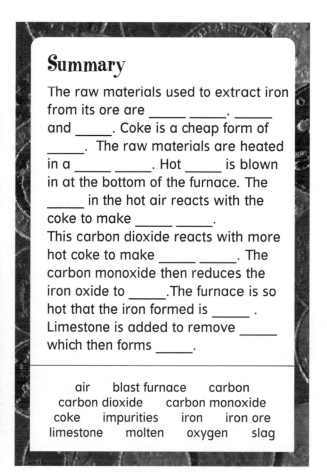

Summary

The raw materials used to extract iron from its ore are _____ _____, _____ and _____. Coke is a cheap form of _____. The raw materials are heated in a _____ _____. Hot _____ is blown in at the bottom of the furnace. The _____ in the hot air reacts with the coke to make _____ _____.
This carbon dioxide reacts with more hot coke to make _____ _____. The carbon monoxide then reduces the iron oxide to _____. The furnace is so hot that the iron formed is _____.
Limestone is added to remove _____ which then forms _____.

air   blast furnace   carbon
carbon dioxide   carbon monoxide
coke   impurities   iron   iron ore
limestone   molten   oxygen   slag

4 When the iron is formed, it is liquid (molten). Why is this?

5 Give two uses for the slag.

! A modern blast furnace can make 3000 tonnes of iron each day using 3000 tonnes of coke and 4000 tonnes of air!

6 Why is the furnace called a blast furnace?

7 Copy and complete the word equation for the first reaction:

carbon + oxygen $\longrightarrow$ _____ _____

8 Most of the iron is used to make steel. Name three useful things that are made from steel.

Ions and electrolytes

What are ions and electrolytes?

Breaking down (**decomposing**) a chemical by using electricity is called **electrolysis**. Very reactive metals like aluminium are higher than carbon in the Reactivity Series. They can't be extracted from their ores by heating with carbon. Instead, we have to use electrolysis.

Many compounds are made up of tiny particles called **ions**. Ions are particles that have an electric charge. They can have either a positive charge (+) or a negative charge (–). For example, sodium chloride (common salt) is made up of sodium ions and chloride ions.

Metals form positive ions and non-metals form negative ions. Sodium ions have a positive charge (+). Chloride ions have a negative charge (–).

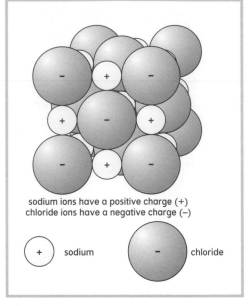

sodium ions have a positive charge (+)
chloride ions have a negative charge (–)

+ sodium   – chloride

A Sodium chloride.

? 1 What is electrolysis?
2 What are ions?

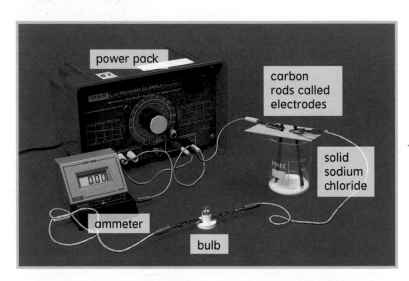

power pack

carbon rods called electrodes

solid sodium chloride

ammeter

bulb

B

C

this can be shown as a circuit diagram

4V

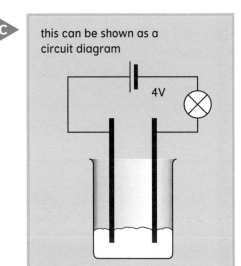

D

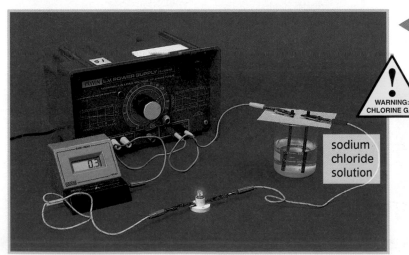

WARNING: CHLORINE GAS

sodium chloride solution

Look at the circuit shown in picture B and diagram C. The bulb is not lit. The solid does not conduct electricity.

If we add water and stir the sodium chloride so that it dissolves and makes a solution, the bulb now lights! The solution conducts electricity.

We can carry out a similar experiment on solid lead bromide. Look at diagram E. The bulb is not lit. Solid lead bromide does not conduct electricity.

Now look at diagram F. The lead bromide has been heated so that it melts. It is now a liquid. The bulb now lights. Molten lead bromide conducts electricity.

If the solid compound is made of ions it won't conduct electricity when it is solid. However, if it is made into a liquid by dissolving it in water or melting it, then it can conduct electricity. Liquids that conduct electricity are called **electrolytes**.

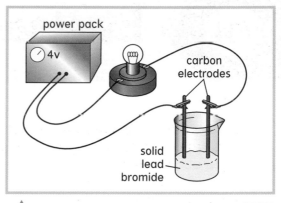

E

F

3 What is an electrolyte?

4 A solid made of ions does not conduct electricity. What two things can we do to make it conduct?

5 What are the carbon rods that dip into the electrolyte called?

Electrolytes and electricity

If ions can move then they can conduct electricity. In a solid, the ions can't move about because they are tightly packed together. But if the solid is dissolved to make a solution, the ions break away from each other and can move about in the water.

Also, if the solid is melted into a liquid, the ions are free to move about.

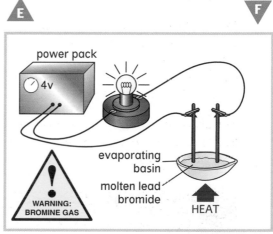

6 Why can't a solid compound made of ions conduct electricity?

7 Why can a compound made of ions conduct electricity after it has been melted?

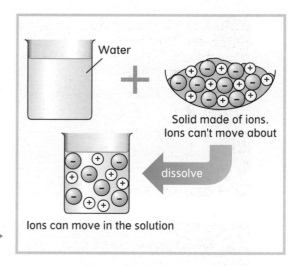

Ions can move in the solution

G

Summary

Compounds that are made of ions can be split up by passing an _____ current through them. This is called _____. Very _____ metals are extracted from their ores by electrolysis. A solid made of ions will not conduct electricity unless it is _____ in water or _____. A liquid that conducts electricity is called an _____. An electrolyte is able to conduct electricity because the _____ can move about.

dissolved   electric   electrolysis   electrolyte
        melted   ions   reactive

8 a) Copper sulphate is a solid made of ions. Will it conduct electricity?

b) What could you do to the solid copper sulphate to make it conduct?

c) Why will it now conduct electricity?

9 Name two metals that can be extracted using electrolysis.

Electrolysis of copper chloride

C14

How does electrolysis actually work?

To understand how we extract reactive metals from their ores, we need to know what happens to the ions in **electrolysis**. Electrolysis is when a compound is split up by electricity.

The electrolysis of copper chloride solution helps us to understand what is going on. Copper chloride is dissolved in water to make a solution and then a circuit is set up.

Look at diagram A. The carbon electrodes have special names. The one connected to the positive (+) terminal on the power pack or battery is called the **anode**. The anode has a positive charge. The electrode connected to the negative (–) terminal is called the **cathode**. The cathode has a negative charge.

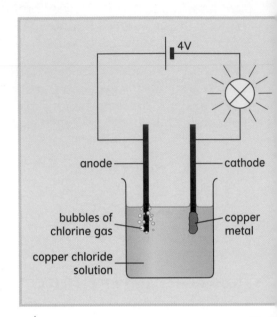

4V

anode — cathode

bubbles of chlorine gas — copper metal

copper chloride solution

A

B

1 What is the name of the electrode that has a positive charge?

2 What is the name of the electrode that has a negative charge?

Copper metal is formed at the cathode (negative electrode).

Chlorine gas is formed at the anode (positive electrode).

The copper chloride has been broken down into its elements, copper and chlorine, by electrolysis.

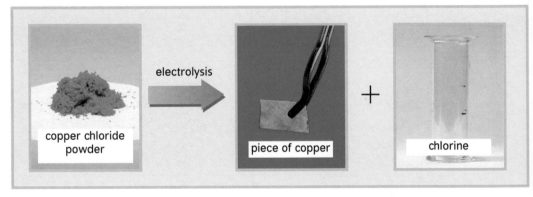

electrolysis

copper chloride powder

piece of copper

+

chlorine

Remember that some compounds like copper chloride are made up of ions. When the compound was formed in the first place, copper atoms and chlorine atoms joined up with each other. When they did this they became **ions** with an electric charge.

When we split up these ions using electrolysis, they turn back into **atoms**.

Malachite is an ore containing copper compounds. How can you use electrolysis to get copper from malachite?

C

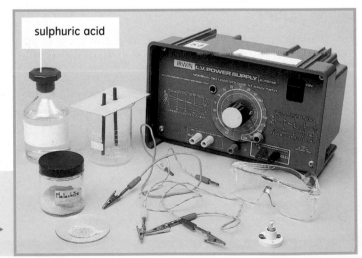

sulphuric acid

Malachite

112

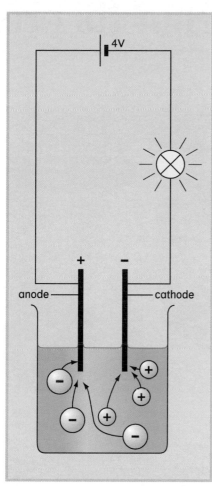

 D

Copper and the cathode

Look carefully at diagram D.

Copper ions have a positive charge. Positive copper ions move to the negative electrode (cathode). The copper ions move to the cathode because they are attracted to it. When they get there the copper ions turn into copper atoms.

Negative ions are attracted to the positive electrode (anode). A gas is usually given off. In this experiment, chlorine gas is given off at the anode.

3 **a)** What charge do copper ions have?
  b) Which electrode do copper ions travel to?
  c) Why do they travel to that electrode?
  d) What happens to copper ions when they get to the cathode?

4 What is the difference between an ion and an atom?

Positive and negative charges attract each other. We usually find that metals are formed at the cathode and non-metals are formed at the anode.

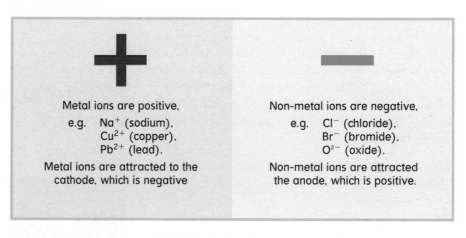

Metal ions are positive,
e.g. Na$^+$ (sodium),
     Cu^{2+} (copper),
     Pb^{2+} (lead).
Metal ions are attracted to the cathode, which is negative

Non-metal ions are negative,
e.g. Cl$^-$ (chloride),
     Br$^-$ (bromide),
     O^{2-} (oxide).
Non-metal ions are attracted the anode, which is positive.

5 **a)** What charge does the cathode have?
  b) What charge does the anode have?

6 **a)** Which electrode do positive ions travel to?
  b) Which electrode do negative ions travel to?

7 If we carried out electrolysis on lead bromide:
  a) what would be formed at the cathode?
  b) what would be formed at the anode?

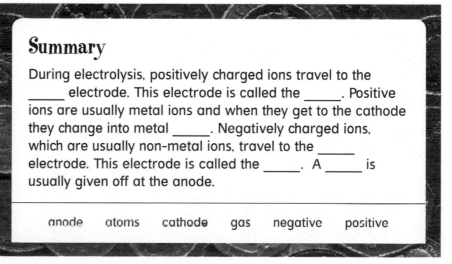

Summary

During electrolysis, positively charged ions travel to the _____ electrode. This electrode is called the _____. Positive ions are usually metal ions and when they get to the cathode they change into metal _____. Negatively charged ions, which are usually non-metal ions, travel to the _____ electrode. This electrode is called the _____. A _____ is usually given off at the anode.

anode   atoms   cathode   gas   negative   positive

Getting aluminium from its ore

How do we get aluminium from its ore?

Aluminium is an extremely useful metal. It is used to make overhead power cables, drink cans, saucepans, aeroplanes, kitchen foil and many other things.

Aluminium is more reactive than carbon. This means that we can't extract it from its ore by heating it with carbon. We have to use **electrolysis**.

?
1 Name two things aluminium can be used for.
2 Why can't we extract aluminium by heating the ore with carbon?

A

Aluminium is found in an ore called **bauxite** (pronounced *'borx –ite'*). It is found near the surface of the ground in Australia, Jamaica, Brazil and other countries.

Digging out bauxite. B

The ore mainly contains aluminium oxide, mixed with impurities like bits of rock and other compounds. The ore must be purified, which means the impurities are taken out of the ore leaving just the aluminium oxide behind.

Then the aluminium can be extracted from the aluminium oxide.

?
3 What is the name of the ore that aluminium is found in?

bauxite (impure aluminium oxide)	→ purify →	**pure aluminium oxide**	→ electrolysis →	**aluminium**

C

? 4 What compound does bauxite mostly contain?

There is more aluminium in the Earth than any other metal, but it was only first extracted in 1827. It was so rare that the Tsar of Russia gave his baby son an aluminium rattle to play with because aluminium was more expensive than gold!

Electrolysis of aluminium oxide

Once we have pure aluminium oxide it has to be melted before electrolysis is carried out. The melting point of aluminium oxide is 2050 °C! This makes it difficult to melt. However, it can be dissolved in a molten chemical called **cryolite** at a much lower temperature, about 850 °C. This saves a lot of energy, but a lot of electricity is still needed to carry out the electrolysis.

In the school laboratory, electrolysis is carried out in a beaker. In industry it is carried out in large containers called **cells**. The one shown in diagram D is about half the size of a classroom.

The oxygen gas reacts with the carbon anodes to make carbon dioxide. The anodes have to be replaced quite often.

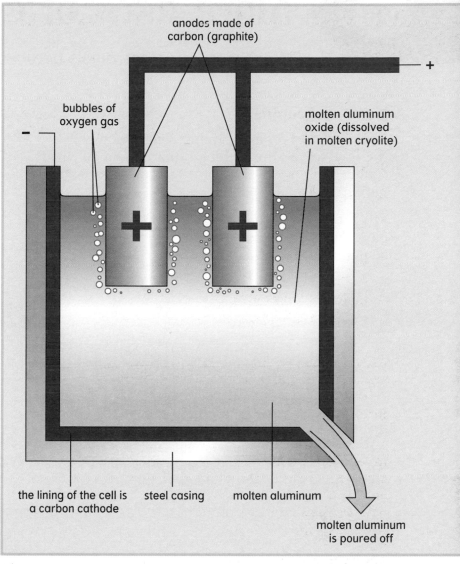

anodes made of carbon (graphite)

bubbles of oxygen gas

molten aluminum oxide (dissolved in molten cryolite)

the lining of the cell is a carbon cathode

steel casing

molten aluminum

molten aluminum is poured off

D

5 What is formed at the cathode?

6 What is formed at the anode?

7 Why do the carbon anodes need to be replaced?

8 What is the raw material used to make aluminium?

9 What must be done to aluminium oxide before electrolysis can take place?

10 Why must it be melted before electrolysis can take place?

11 Why do you think it is expensive to extract aluminium from its ore?

Summary

The raw material for making aluminium is purified _____ _____ (bauxite). Because aluminium oxide has a very high melting point, it is first _____ in molten _____ at a much _____ temperature, which saves energy. Aluminium is formed at the _____ and oxygen is formed at the _____. The oxygen reacts with the carbon _____ and so they have to be replaced quite often.

aluminium oxide	anode	anodes	cathode
cryolite	dissolved	lower	

More uses of electrolysis

What else can electrolysis be used for?

Purifying copper

Copper is used for many things. One reason why it is used for electrical wiring is that it is an excellent conductor of electricity. However, it must be very **pure** to do its job well.

When copper is extracted from its ore, it is not pure enough to be used for electrical wiring. It can be purified using electrolysis.

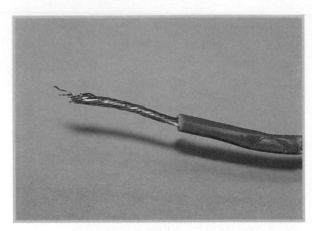

 1 Why is copper used for electrical wiring?

2 Why must it be pure?

Diagram B shows how electrolysis can be used to purify copper. It can be done this way in a school laboratory.

A

When purifying copper, the sludge that is left over is not wasted. It contains precious metals like silver and gold.

In industry, electrolysis is carried out in large cells (like the extraction of aluminium) and there are many cells running at the same time. Diagram B shows the electrolysis before and after. In the laboratory, you can see the results in about 15 minutes.

B

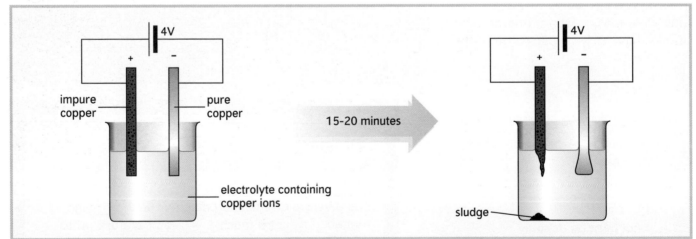

The anode (positive electrode) is made out of the impure copper that you are trying to purify. The cathode is made out of pure copper. The electrolyte contains copper ions.

The anode (impure copper) gets smaller and smaller. The cathode (pure copper) gets bigger and bigger. The impurities fall off into the solution forming sludge at the bottom of the beaker.

 3 What is the cathode made from?

4 What is the anode made from?

5 What must the electrolyte contain?

Electroplating

Look at diagram D. If you swap the cathode for a metal object like a key, then the key becomes coated with copper.

You can coat lots of metal objects with a different metal using electrolysis. This is called **electroplating**.

If you have a bike, the handlebars might be covered with chromium (chrome). This is a very shiny metal which does not rust. Underneath, the handlebars are actually made of steel. The steel has a layer of chromium to protect it from rusting. It is put onto the steel by electroplating.

Many metal objects are electroplated with different metals to protect them from rusting or to make them look nice. For example, most silver cups and trophies are really only steel with a thin layer of silver on them.

? 6 What is electroplating?

7 Name two objects that are often electroplated with another metal.

P How would you investigate what affects the speed of electrolysis?

copper sulphate solution

C

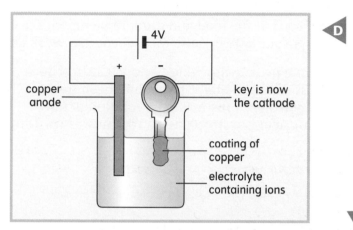

D

4V

+   −

copper anode

key is now the cathode

coating of copper

electrolyte containing ions

E

Summary

Copper is an excellent _____ of electricity and is used for _____ wiring. To do this job well, it must be very _____. _____ is used to purify copper. The anode is made of _____ copper and the cathode is made of _____ _____. The electrolyte must contain a solution of copper _____. The anode gets _____ and the cathode gets bigger. Covering a metal with another using electrolysis is called _____.

conductor	electrical	electrolysis
electroplating	impure	ions	pure
pure copper	smaller

? 8 a) What happens to the cathode when copper is being purified?
   b) What happens to the anode?

9 What happens to the impurities?

10 Describe how you could cover a small spoon in a layer of copper.

11 You cannot electroplate a plastic spoon. Explain why electroplating will not work on plastic spoons.

Acids and alkalis

What are acids and alkalis?

Some compounds dissolve in water. When they do, an **aqueous solution** is made.

Aqueous just means 'dissolved in water'. The solution can be an **acid**, an **alkali** or it can be **neutral**.

When you think of the word acid you might think of something dangerous like battery acid. But not all acids are dangerous. Look at the food and drink in picture A. They all have a sharp, sour taste because they all contain **weak acids**.

You may have used hydrochloric acid (HCl) in school. When it dissolves in water it forms hydrogen ions (H^+). The **hydrogen ions** make the solution **acidic**. Hydrochloric acid is a **strong acid**. You must *never* drink it!

? 1 Write down one property of an acid.

2 Name two foods or drinks that contain weak acids.

A *All these foods and drinks contain acids.*

Picture B shows some substances that contain alkalis. Some of them are **weak alkalis**, like soap and toothpaste. They are safe to use. But, some are **strong alkalis** and so are dangerous, like oven cleaner.

A strong alkali you may have used in school is sodium hydroxide (NaOH). When it dissolves it forms hydroxide ions (OH^-). The hydroxide ions make the solution alkaline.

Neutral solutions

Acids and alkalis are chemical opposites, a bit like up and down or black and white.

Some solutions are neither acid nor alkali. They are **neutral**.

It's a bit like a football match. Somebody might watch the game as a *neutral supporter*. They don't support either team.

Pure water is neutral. It is not an acid or an alkali.

? 3 Name a material that contains a weak alkali.

4 Name a strong alkali.

5 Name a neutral substance.

B *All of these contain alkalis.*

Indicators

Indicators tell us if a chemical is an acid, an alkali or is neutral. **Litmus** is an indicator. It can be in a liquid form or on paper.

Acids turn litmus indicator **red**.

Alkalis turn litmus indicator **blue**.

Neutral solutions don't change the colour of litmus indicator.

Litmus indicator does not tell us how strong an acid or alkali is.

Universal indicator turns different colours depending on how strong the acid or alkali is. The different colours are put onto a scale and given a number called a pH number. This is the pH scale.

 Litmus indicator comes in different forms.

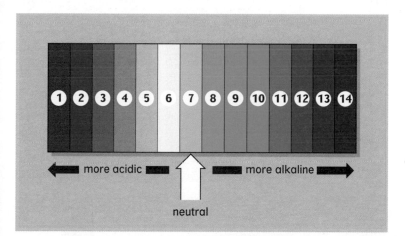

more acidic ← → more alkaline

neutral

You can see that the lower the pH number, the stronger the acid. The higher the pH number, the stronger the alkali.

A neutral solution is pH 7.

 The pH scale

Summary

When compounds are dissolved in water, the solution can be an _____, an alkali or _____. Litmus indicator turns _____ in acids and _____ in alkalis. Universal indicator tells us the _____ of a solution. _____ have pH numbers between 1 and 6. Alkalis have pH numbers between _____ and a neutral solution has a pH number of _____. The _____ the pH number, the stronger the acid and the _____ the pH number, the stronger the alkali.

7	8 and 14	acid	acids	blue	higher
	lower	neutral	pH	red	

6 Which pH number is the strongest acid?

7 Which pH number is the strongest alkali?

8 What is an aqueous solution?

9 What pH is water?

10 Match the pH numbers with the solutions.

pH: 3, 6, 7, 9, 14

solutions: Fizzy orange, oven cleaner, indigestion tablets, distilled water, milk

Neutralisation

What happens when an acid and an alkali react together?

Acids and alkalis are chemical opposites. If we mix them together in the right amount, they react together and cancel each other out. A **salt** and **water** are made.

The salt is dissolved in the water and forms a **neutral solution**, which is pH 7. An indicator such as universal indicator can show that this has happened.

The word equation is:

acid + alkali ⟶ salt + water

When an acid and an alkali react together to make salt and water, the reaction is called **neutralisation**.

1 What is neutralisation?

2 What pH is a neutral solution?

You may have had indigestion or heartburn. The burning feeling you get is caused by too much acid in your stomach. You can neutralise the acid by taking an antacid tablet. This contains a weak alkali which reacts with the acid.

A bee's sting is acidic. It can be neutralised with an alkali. Bicarbonate of soda (a weak alkali) can be used to ease the pain.

A wasp's sting is alkali. Vinegar can be used to relieve the pain of a wasp sting because the vinegar is an acid and it neutralises the alkali in the wasp's sting.

3 Why can vinegar be used on a wasp's sting?

A *These tablets neutralise excess acid in your stomach.*

 B

bee sting neutralised by

wasp sting neutralised by

Salts

We know that a salt is made in neutralisation. Most salts contain metals. Look at this word equation:

hydrochloric acid + sodium hydroxide ⟶ sodium chloride + water

HCl + $NaOH$ ⟶ $NaCl$ + H_2O

an acid + an alkali ⟶ a salt + water

Notice that the metal part of the salt formed is sodium and that it came from the alkali. The salt is called sodium chloride and is the common salt that is put on your food.

 C

The salt made depends on the *metal* in the alkali. For example, if we used potassium hydroxide instead of sodium hydroxide:

hydrochloric acid + potassium hydroxide ➔ potassium chloride + water

    HCl   +   KOH   ➔   KCl   +   H₂0

This time the salt is called potassium chloride.

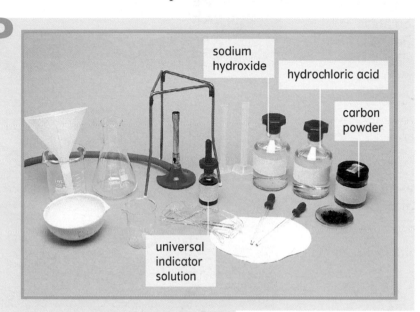

sodium hydroxide
hydrochloric acid
carbon powder
universal indicator solution

D How could you make sodium chloride?

The salt made also depends on which acid is used. Different acids make different salts. Table E shows which type of salt is made from different acids.

4 What two things does the type of salt made depend on?

5 What type of salt is made if sulphuric acid is used?

Acid	Type of salt made	Example
hydrochloric acid HCl	chlorides	sodium chloride NaCl
sulphuric acid H₂SO₄	sulphates	magnesium sulphate MgSO₄
nitric acid HNO₃	nitrates	potassium nitrate KNO₃

Summary

An acid and an _____ react together to make salt and _____, the reaction is called _____. The word equation is:

_____ + _____ ➔ salt + water.

The salt made depends on two things: the _____ in the alkali and the type of _____ used. Hydrochloric acid produces _____ salts, nitric acid produces _____ salts and _____ acid produces _____ salts.

acid   alkali   chloride
metal   neutralisation   nitrate
sulphate   sulphuric   water

6 Copy and complete this word equation:

_____ + alkali ⟶ _____ + water

7 How could you neutralise an acid?

8 How could you neutralise an alkali?

9 If hydrochloric acid were reacted with calcium hydroxide, what would the salt be called?

10 You have 25 cm³ of acid in a beaker. The pH is 1.
  a) Describe how the pH of the acid would change as you slowly added 30 cm³ of alkali.
  b) Explain why these changes happen.

Transition metal salts

How can we make transition metal salts?

We have seen how salts are made by neutralisation. Salts are made by reacting an acid with an alkali. Alkalis are part of a larger group of compounds called **bases**. Bases that dissolve in water are called alkalis and form **alkaline solutions**. Most bases do not dissolve in water, but they still react with acids.

 1 How are alkalis different from other bases?

Transition metal salts can be made by reacting transition metal oxides or hydroxides with acids.

Transition metal oxides and hydroxides are bases but do not dissolve in water. However, they will react with acids.

The general equation is:

acid　　　+　　base　　⟶　　　salt　　+ water

e.g. sulphuric acid + zinc oxide ⟶ zinc sulphate + water

$$H_2SO_4 \quad + \quad ZnO \quad \longrightarrow \quad ZnSO_4 \quad + \quad H_2O$$

A *Transition metal salts can be very useful. For example, copper sulphate can be used in a spray to protect these grapevines from insects.*

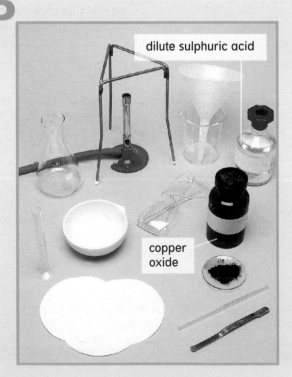

dilute sulphuric acid

copper oxide

B How would you make copper sulphate (a transition metal salt)?

Non-metal salts

Ammonia (NH_3) is a compound made from two non-metals, nitrogen and hydrogen. Ammonia is a gas and dissolves in water to form ammonia solution. Ammonia is an alkali. It can be neutralised by reacting it with an acid to make an ammonium salt. For example:

ammonia + sulphuric acid ⟶ ammonium sulphate

$$2\,NH_3 \quad + \quad H_2SO_4 \quad \longrightarrow \quad (NH_4)_2SO_4$$

Ammonium sulphate can be used in fertilisers.

Another ammonium salt used in fertilisers is ammonium nitrate. It can be made by neutralising ammonia solution with nitric acid.

ammonia + nitric acid ⟶ ammonium nitrate

$$NH_3 \quad + \quad HNO_3 \quad\quad\quad NH_4NO_3$$

2 What type of salt is made when ammonia solution reacts with an acid?

3 Name a salt that can be used to make fertilisers.

P How would you make a salt used in fertilisers?

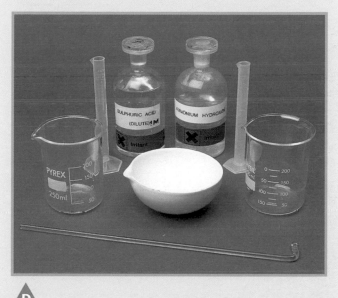

C

D

Summary

Bases that dissolve in water are called _____ and form alkaline solutions. Transition metal salts can be made by reacting an acid with a transition metal _____ or _____. Ammonia is a gas that dissolves in water to form an _____ solution. This can be neutralised with an acid to make _____ salts.

---

alkaline   alkalis   ammonium
hydroxide   oxide

4 Copy and complete table E to name the transition metal salts made when different acids and bases are reacted together.

E

Acid	Base	Salt
sulphuric acid	zinc oxide	zinc sulphate
nitric acid	copper oxide	
hydrochloric acid	copper hydroxide	

5 a) Write a word equation to show how ammonium chloride can be formed.
   b) Write a symbol equation for the reaction.

123

Further questions

1 Copy and complete these sentences using words from the box. You may use each word once, more than once, or not at all.

Metals like copper and aluminium are good conductors of _____. This means that copper can used to make _____ and aluminium can be used to make overhead _____ _____. Iron is used to make _____ and cars because it is very _____. (5)

bridges	power cables	electricity
energy	flexible	nails
strong	weak	wires

2 The diagram shows part of the Periodic Table.

sodium	magnesium		aluminium	silicon	phosphorous	sulphur	chlorine	argon
11	12		13	14	15	16	17	18

a) From the elements in the table, write down the name of

i) a group 1 metal
ii) a group 2 metal
iii) a non-metal. (3)

b) What is the name of the block of metals that is in the centre of the Periodic Table? (1)

3 This question is about acids and alkalis.

a) Copy and complete the word equation for neutralisation.

acid + alkali ⟶ _____ + _____ (2)

b) Name the salt that would be formed by reacting dilute hydrochloric acid with sodium hydroxide. (1)

c) Name the acid and alkali needed to make the salt sodium nitrate. (2)

4 Group 1 metals and transition metals have similar properties, but there are also some differences. The table shows some properties of group 1 metals and transition metals. Two rows have been completed for you. Write down the letters of the other boxes that should have ticks in them. (5)

Property	Group 1 metal	Transition metal
shiny	✓	✓
hard	✗	✓
soft	a)	b)
good conductor of electricity	c)	d)
good conductor of heat	e)	f)
low density	g)	h)
high density	i)	j)

5 a) Copy and complete these sentences using words from the box. You may use each word once, more than once, or not at all.

When metals react with oxygen from the air, a metal _____ is made.

When reactive metals like sodium react with water a metal _____ is made. _____ gas is also made.

When metals react with dilute acids, a metal _____ and hydrogen gas are made. (4)

carbonate	gas	hydrogen	hydroxide
nitrogen	oxide	salt	sulphides

b) When sodium reacts with water an alkaline solution is made. What colour would litmus indicator turn in this solution? (1)

6 Three metals, X, Y and Z, were found on another planet. The table shows their reactions with water and dilute acid.

Metal	Reaction with water	Reaction with acid
X	no reaction	slow reaction
Y	fast reaction	violent reaction
Z	no reation	no reaction

a) Arrange the three metals in order of reactivity starting with the most reactive. (1)

b) Another test was tried on metal X, as shown in the diagram below.

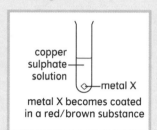

copper sulphate solution

metal X

metal X becomes coated in a red/brown substance

Explain, as fully as you can, the results of this test. (2)

7 Look at the diagram below.

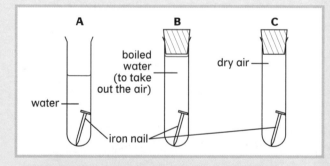

A

B

C

boiled water (to take out the air)

dry air

water

iron nail

a) Explain, as fully as you can, why the nails in tubes B and C do not rust. (2)

b) Write down two ways that rusting can be prevented. (2)

c) The picture below shows the hull of a ship. The zinc blocks protect the hull from rusting. Use the Reactivity Series to help you explain why the zinc blocks corrode away instead of the iron hull. (2)

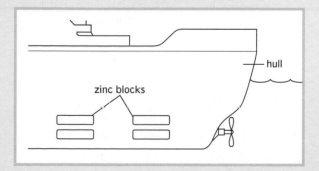

hull

zinc blocks

8 Some metals are found in a native state. However, most metals are found as metal compounds mixed up with rock. This mixture is called an ore.

a) Explain what native means. (1)

b) Name a metal that is found in a native state. (1)

c) Name a metal that is found as an ore. (1)

d) Why are some metals found as ores and others found in a native state? (2)

9 Look the Reactivity Series below.

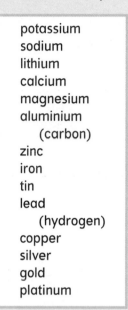

potassium
sodium
lithium
calcium
magnesium
aluminium
   (carbon)
zinc
iron
tin
lead
   (hydrogen)
copper
silver
gold
platinum

a) i) Write down the name of a metal that could be extracted by heating its ore with carbon. (1)

ii) Explain your answer (1)

b) i) Write down the name of a metal that cannot be extracted by heating its ore with carbon. (1)

ii) Explain your answer (1)

iii) How could you extract this metal from its ore? (1)

10 A student was trying to extract the metals from lead oxide and aluminium oxide. She heated each oxide with carbon in a fume cupboard. She was able to extract lead from lead oxide but not aluminium from aluminium oxide.

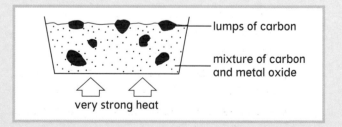

lumps of carbon

mixture of carbon and metal oxide

very strong heat

a) Explain as fully as you can, the results of these experiments. (2)

b) i) Copy and complete the word equation for the reaction between lead oxide and carbon.

lead oxide + carbon ⟶ ____ + ____

ii) What is reduced in this reaction?

iii) What is oxidised in this reaction? (4)

The planet Earth, our home

What is the Earth made from?

A

Scientists think that the Earth was formed 4.6 billion years ago from a ball of gas and dust. Our planet is still changing now. While you are reading this volcanoes are erupting, mountains are being made, earthquakes are shaking the ground and the surface you are standing on is moving!

The Earth

The Earth is made up of three major layers. The outer layer is called the **crust** and it is made of solid rock. Most of the crust is hidden by soil or water but in some places the rock can be seen. The layer under the crust is called the **mantle**. Although the rock in the mantle is almost completly solid, it is so hot that it can flow, a bit like very thick treacle. The outer part of the Earth is moved by the flowing mantle all the time. The middle of the Earth is called the **core**.

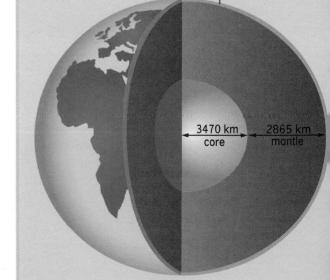

the crust has an average thickness of 35 km

3470 km core

2865 km mantle

B

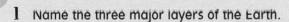
1 Name the three major layers of the Earth.

2 What is the average thickness of the crust?

3 What is the mantle made of?

4 How many kilometres thick is the mantle?

5 What makes the outer part of the Earth move?

Two thirds of the Earth's crust is under the sea.

The core of the Earth is made of the metals iron and nickel, and it is very hot. The inner part of the core is solid but the outer part is a viscous (very thick) liquid a bit like the mantle.

6 What two metals are found in the core?

7 What is the radius of the core?

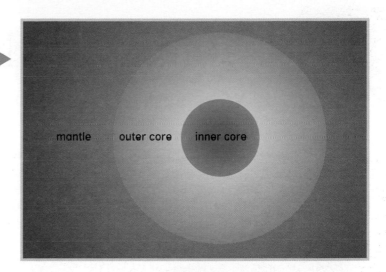
mantle   outer core   inner core

Density

The density of a material tells you what the mass of 1 cm³ of it would be. Lead is a dense material. A 1 cm³ block of lead has a big mass and it feels heavy. Polystyrene is not very dense at all. A 1 cm³ block of polystyrene has a small mass and it feels light. The molecules in a dense material are packed together very tightly so 1 cm³ of it has a bigger mass.

P How could you work out the density of different rocks?

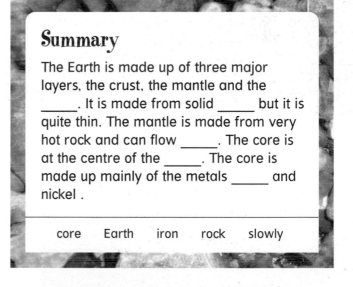
F

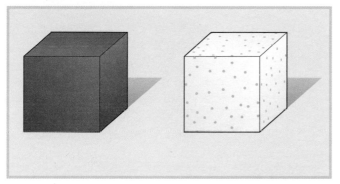

 D 1 cm³ of lead has a mass of 11.35 grams.

 E 1 cm³ of polystyrene has a mass of 0.01grams/cm³.

Scientists have worked out that the overall density of the Earth is much greater than the density of the crust. This means that the core and the mantle must be made from materials that are much denser than the crust. The core is the densest part of the Earth. A 1 cm³ piece of the core would have a much bigger mass and feel a lot heavier than a 1 cm³ piece of the crust.

Summary

The Earth is made up of three major layers, the crust, the mantle and the _____. It is made from solid _____ but it is quite thin. The mantle is made from very hot rock and can flow _____. The core is at the centre of the _____. The core is made up mainly of the metals _____ and nickel .

core   Earth   iron   rock   slowly

8 **a)** Which has the bigger mass, 1 cm³ of lead or 1 cm³ of polystyrene?

   b) Which is denser, lead or polystyrene?

9 **a)** Which is the least dense layer of the Earth?

   b) Which is the most dense layer of the Earth?

10 Which is denser, the Earth's crust or what is inside the Earth (the interior)? How do scientists know this?

11 Find out what a geologist does.

Igneous rocks

What are igneous rocks and how were they formed?

There are many different types of rock which make up the crust. You may have heard of rocks like granite, sandstone, limestone and marble. Rocks like these are classified into three groups depending on how they were made. The three groups are **igneous** rocks, **sedimentary** rocks and **metamorphic** rocks.

 1 Write down the names of four rocks.

Rocks under the surface of the Earth can get so hot that they melt to form a liquid rock called **magma**. The hot magma can rise up in the Earth's crust or out onto the Earth's surface. If this happens the magma cools down and turns into solid rock. Rocks made from magma like this are called **igneous** rocks.

 2 a) What is magma.
b) What does magma form when it cools down.

Magma is less dense than the rocks around it, so it forces its way upwards. If it reaches the surface a **volcano** is formed.

 C

A

When volcanoes erupt red hot magma pours out. Magma that flows down the side of a volcano is called **lava**. When the eruption is over the lava cools down and freezes (turns into a solid) as a layer of igneous rock. Each new eruption adds another layer of rock to the volcano and so it gets bigger.

B

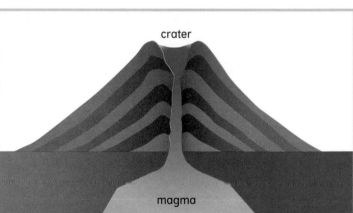

crater

magma

A volcano may not erupt for many years. It is said to be **dormant** (or **inactive**). Some dormant volcanoes suddenly erupt again without warning.

3 What happens when a volcano erupts?

4 What is lava?

Extrusive rocks are formed from magma which comes out of volcanoes (lava) onto the Earth's surface. As lava cools down, it forms crystals. Lava on the Earth's surface cools down very quickly and the rock which is made has small crystals. Rock formed like this is called extrusive igneous rock. Basalt is an extrusive igneous rock.

Intrusive rocks are made from magma which cools down inside the crust. This magma cools more slowly and the rock which is made has large crystals. Rock formed like this is called intrusive igneous rock. Granite is an intrusive igneous rock.

D *Basalt has small crystals.*

F *Granite has big crystals.*

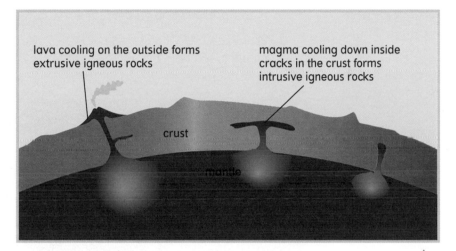

lava cooling on the outside forms extrusive igneous rocks

magma cooling down inside cracks in the crust forms intrusive igneous rocks

crust

mantle

E

5 Write down the name of one extrusive igneous rock.

6 Write down the name of one intrusive igneous rock.

Summary

Igneous rocks were made from hot liquid rock called _____ which cooled down and froze. Magma which flowed out of volcanoes as _____ cooled down quickly on the Earth's surface to make extrusive igneous rock. An extrusive igneous rock like _____ has _____ crystals. Magma which cooled down slowly inside the Earth's crust made intrusive igneous rock. An intrusive igneous rock like _____ has _____ crystals.

basalt   big   granite   lava
magma   small

P How could you investigate the sizes of crystals that form at different temperatures?

G

7 Why does granite have bigger crystals than basalt?

8 Look at diagram C. How many times do you think the volcano has erupted? Write a sentence to explain your answer.

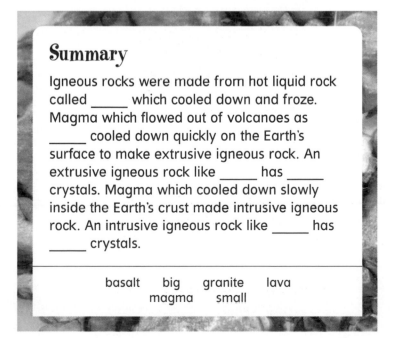

D3 Sedimentary rocks

What are sedimentary rocks and how were they formed?

Although rocks seem very hard, they are being broken into smaller and smaller pieces all the time. Rocks are broken up into smaller pieces by **weathering** and **erosion**.

When they have been **deposited** and buried these particles can be **compressed** (squeezed) and **cemented** (glued) back together to make new rocks called **sedimentary** rocks.

 1 What does weathering and erosion do to rocks?

Flow chart A tells you how sedimentary rocks are made. **A**

| **Weathering and erosion:** Rocks are worn away into tiny particles by the weather, acid rain, wind, flowing water and animals and plants. |

↓

| **Transport:** Weathered and eroded particles of rock are carried away by wind or water. |

↓

| **Deposition:** The water or wind carrying the tiny particles of rock slows down. The particles are **deposited**. This means they fall to the ground or sink to the bottom of the river or the sea. The tiny fragments that settle out are called **sediment**. |

↓

| **Compression and cementation:** Over millions of years the deposited sediment is **compressed** (or squashed) by the weight of more material which settles on top of it. Crystals grow between the grains and cement them together |

↓

| **New rock formed:** The sediment sticks together and hardens, forming new sedimentary rock. |

 2 How are particles of rock transported?

3 What happens to rock particles when they are deposited?

4 What happens to the sediment when it is compressed and cemented?

5 What is made when sediment is compressed and cemented together?

B *Sandstone is made of grains of sand cemented together.*

C *Conglomerate is made of small pebbles cemented together.*

 6 What is sandstone made from?

7 What is conglomerate made from?

 The Mississippi river in America carries about 340 million tonnes of sediment every year.

Limestone is a sedimentary rock which was made from the remains of sea creatures. In this case the sediment was the fragments of shells which settled at the bottom of the sea. They were compressed and cemented when other material settled on top of them. Sea shells are made of calcium carbonate. This is why limestone is made mainly of calcium carbonate.

 8 Why is limestone mainly calcium carbonate?

Sedimentary rocks are usually formed in layers. This is because sediments are not deposited all the time. Each time sediment is deposited, a new layer is formed.

Ripple marks

If you go down to a beach when the tide has just gone out, the sand often looks bumpy. When the tide was in, particles of sand were deposited on the beach. Sometimes currents or waves deposit this sand in patterns, and this is where the bumps come from. This also happened millions of years ago when sedimentary rock was made. When the sediment, became sedimentary rock the patterns became part of the rock. These patterns are called **ripple marks**.

 9 How were ripple marks made?

10 Put these words in order to explain how sedimentary rock is made and explain what each word means.

deposition   rock formation   weathering
transport

11 Explain how limestone was made.

Summary

Sedimentary rocks are made from tiny _____ of other rocks which have been _____ or squeezed and cemented together over millions of years. Different kinds of _____ rock are made from different kinds of rock particle. Sandstone is made from particles of _____. Sedimentary rocks are often found as _____ of rock.

compressed   layers   particles
sand   sedimentary

 D *Limestone rocks.*

Layers of sedimentary rocks. **E**

 Ripples in sand.

Ripple marks in rock. **G**

131

Evidence from rocks

What can rocks tell us about the way they were formed?

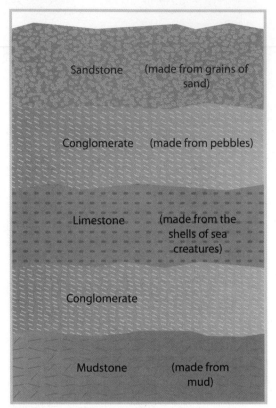

Sandstone (made from grains of sand)

Conglomerate (made from pebbles)

Limestone (made from the shells of sea creatures)

Conglomerate

Mudstone (made from mud)

A

Sedimentary rocks are usually formed in layers. This is because sediments are not deposited all the time. Each time sediment is deposited, a new layer is formed.

The deeper down a layer of sedimentary rock is, then the older it is. Any layers of rock which are on top of it must have been made from sediment which was deposited later so these rocks must be younger.

 Look at diagram A and answer these questions.

1 How many layers of sedimentary rock are there here?

2 How many layers of conglomerate are there?

3 Copy and complete the following sentences.

a) Sandstone is the _____ (oldest/youngest) layer of rock as it is found at the top.

b) Mudstone is the oldest layer of rock as it is found at the _____ (top/bottom).

Fossils

When animals and plants die their bodies usually rot away, or **decay**. Sometimes the remains of a dead animal or plant do not rot and its body is **preserved**. The preserved remains of dead animals and plants are called **fossils**.

When the animals and plants died, their remains fell to the bottom of lakes and rivers along with the sediment. As the sediment was compressed into solid rock, the remains were preserved as fossils.

B

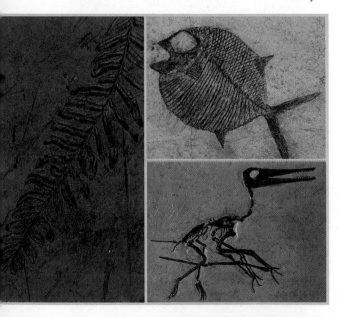

The oldest human fossil was found in Kenya, a country in Africa. It was between 2 and 3 million years old.

 4 What are fossils?

P How could you make your own casts of fossils?

C

A layer of sedimentary rock may contain fossils of plants and animals that lived at the time that the rock was made. This means that if two rock samples from different places have the same kinds of fossils, they are likely to be about the same age.

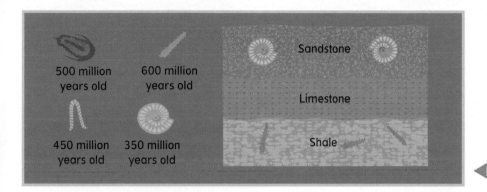

Scientists know how old different kinds of fossils are. If they find a type of fossil in a layer of sedimentary rock it will tell them how old the rock is.

5 Look at the four rock samples in diagram D. Match them up into pairs of the same age.

500 million years old   600 million years old

450 million years old   350 million years old

Sandstone

Limestone

Shale

6 Look at the fossils and layers of rock in diagram E.

a) How old is the layer of shale?

b) How old is the layer of sandstone

7 Look at diagram F. It shows the layers of rock at two places, X and Y.

a) How old is the layer of sandstone?
b) How old is the layer of conglomerate? How do you know this?

Mudstone (450 million years old)

Sandstone (500 million years old)

place X

Conglomerate

Mudstone

Limestone

place Y

The layer of mudstone at place Y is above the limestone, so it must have been made after the limestone. The mudstone is below the conglomerate, so it must have been made before the conglomerate. The sediment which made the mudstone must have been deposited between 450 and 500 million years ago.

Scientists say that the sedimentary rock at place X was made by **discontinuous deposition**. This means that sediment was not being deposited all the time. Sediment was deposited 500 million years ago to make the sandstone. Then there was no sediment deposited for the next 50 million years. More sediment was then deposited 450 million years ago to make the conglomerate.

8 How do scientists know that no sediment was deposited at place X between 450 and 500 millions years ago?

9 What is discontinuous deposition?

10 The thickness of a sediment layer can change. What things might affect the thickness of sediment layers?

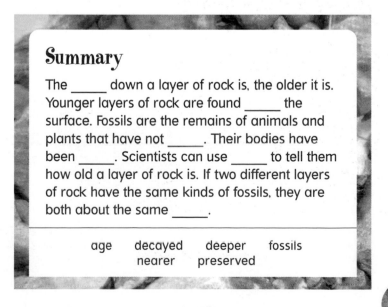

Summary

The _____ down a layer of rock is, the older it is. Younger layers of rock are found _____ the surface. Fossils are the remains of animals and plants that have not _____. Their bodies have been _____. Scientists can use _____ to tell them how old a layer of rock is. If two different layers of rock have the same kinds of fossils, they are both about the same _____.

age	decayed	deeper	fossils
	nearer	preserved	

Snapping and bending

How do rocks in the Earth's crust get bent and broken?

Some scientists examined the fossils that they found in the rocks in diagram A. They worked out how old the fossils were.

 1 a) Which fossil do you think was the oldest?
  b) Why do you think so?

The scientists thought that fossil X was younger than fossil Y because it was found in a layer of rock that was nearer the surface.

They were surprised to find out that fossil X was actually older than fossil Y. They worked out that movements in the crust must have turned the layers of rock upside down.

When sediment sank to the sea bed or the river bed it formed a horizontal (flat) layer. When the sediment was compressed, it formed horizontal layers of sedimentary rock.

Many layers of sedimentary rock are now found bent or broken. This shows that the Earth's crust must have moved.

 2 What were layers of sedimentary rock like when they were first made?

Folds

When the Earth's surface moves huge forces can push or pull layers of rock in the crust. Sometimes these forces can bend layers of rocks into **folds**.

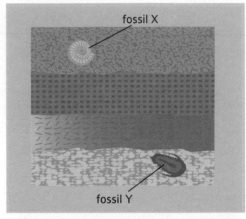

A *Sediment sinks to the sea bed.* **B**

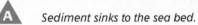

 3 What is a fold?

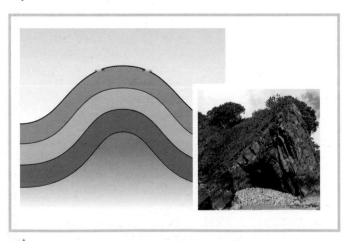

C *Layers of rock can be folded upwards.*

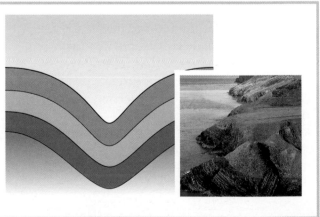

Layers of rock can be folded downwards. **D**

 134

P How would you show the effects of movements in the Earth's crust? **E**

Faults

A sudden movement in the crust can break rock. This causes a **fault**. It is easy to see a fault when sedimentary rock snaps as the layers don't line up anymore.

4 What is a fault?

5 How can you tell there is a fault in picture F?

F

6 Look at the four drawings, G to J. They show how some rocks changed over millions of years.

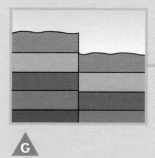 G

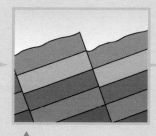 H

 I

J

Write these sentences out in the correct order to explain what has happened to the rocks.

- The tilted layers of rock have been weathered and worn away.

- Movement in the crust has snapped and faulted the layers of rock.

- More sediment has been deposited to make another layer of sedimentary rock.

- The faulted layers of rock have been tilted.

7 Were the layers of rock in drawings G to J made by continuous or discontinuous deposition? Explain your answer.

Summary

The Earth's surface is _____ all the time. This puts great forces on the rocks in the _____. When sedimentary rock was made it was laid down flat as a horizontal _____ when sediment was deposited and then _____. It is easy to see how movements have shaped the Earth's crust by looking at what has happened to layers of _____ rock. Sometimes rocks in the crust are squeezed up or down into _____. Sudden movements of the Earth's surface can snap rock and cause _____.

compressed   crust   faults   folds   layer
moving   sedimentary

Metamorphic rocks

What are metamorphic rocks and how were they formed?

Ever since the Earth was formed the crust has continued to move. Sometimes when the crust buckles and folds, igneous and sedimentary rocks which were on the surface can get buried deep underground. When this happens the rocks can be changed into new kinds of rocks called **metamorphic** rocks.

If rocks are pushed underground they are compressed (squashed) by huge pressures and heated. The temperatures are not hot enough to melt the rock back into magma but the pressure and heat can change the structure of the rock.

Rocks contain chemicals called **minerals**. Often these minerals are randomly arranged with no real pattern. Huge pressures and high temperatures can force these minerals to line up in **layers**. When this happens the structure of the rock changes and it becomes a new kind of rock called a metamorphic rock.

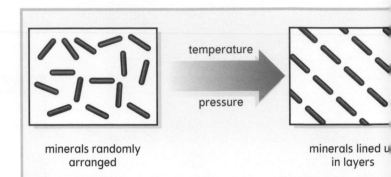

temperature

pressure

minerals randomly
arranged

minerals lined u
in layers

 A

 1 How are igneous and sedimentary rocks turned into metamorphic rock?

2 Why do many metamorphic rocks have layers?

3 Copy and complete the following sentences.

a) Marble, quartzite and slate are all _____. rocks.

b) Marble was made from _____.
Slate was made from _____ and quartzite was made from _____.

 B

Original rock

Limestone

Sandstone

Mudstone

Metamorphic rock

Marble

Quartzite

Slate

Many mountains are made from metamorphic rock. The mountains were made by movements in the crust which pushed layers of rock upwards.

The rock which was pushed and squeezed upwards was put under huge pressure. This pressure compressed the rock and changed it into metamorphic rock. Map C shows mountain ranges which are made from metamorphic rock.

4 Why are many mountains made of metamorphic rock?

5 Which mountains in Britain are made of metamorphic rock?

Summary

Igneous and sedimentary rocks can be turned into _____ rocks if they are compressed under very high temperatures and _____. Metamorphic rocks have a different structure from the rocks they were made from. Many have layers because the _____ and pressure have forced the _____ to line up in layers. Limestone is a _____ rock which can be turned into the metamorphic rock _____. Sandstone is a _____ rock which can be turned into the metamorphic rock _____. Many _____ belts are made from metamorphic rock.

marble	metamorphic	
minerals	mountain	
pressures	quartzite	
sedimentary	temperature	

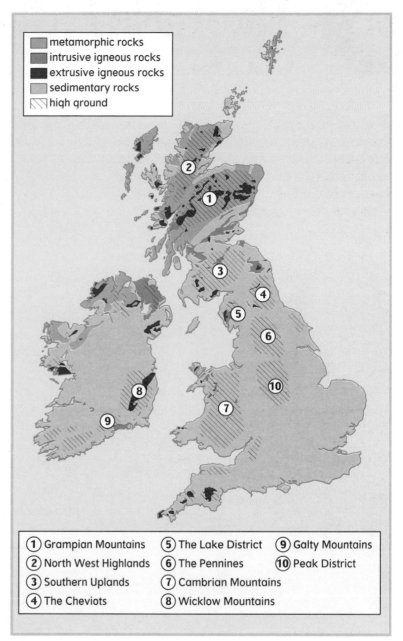

metamorphic rocks		
intrusive igneous rocks		
extrusive igneous rocks		
sedimentary rocks		
high ground		

① Grampian Mountains   ⑤ The Lake District   ⑨ Galty Mountains
② North West Highlands   ⑥ The Pennines   ⑩ Peak District
③ Southern Uplands   ⑦ Cambrian Mountains
④ The Cheviots   ⑧ Wicklow Mountains

C

6 Copy and complete table D. Put each of the following rocks in the correct column.

sandstone   basalt   limestone   granite   marble
mudstone   slate   conglomerate

D

Igneous	Sedimentary	Metamorphic

7 Find out what the metamorphic rocks slate and marble are commonly used for.

The rock cycle

What is the rock cycle?

Rocks are being worn away all the time by weathering and erosion. At the same time sediments are being deposited and new sedimentary rocks are being made. From time to time melting occurs and new igneous rocks form, or rocks are compressed and heated, forming new metamorphic rocks. All these processes are linked together in the **rock cycle**

Diagram A shows you what happens in the rock cycle.

A

3 Copy and complete the following sentences.

  a) Magma which cools down inside the Earth's crust forms _____ igneous rock.

  b) Magma which cools slowly forms igneous rock with _____ crystals.

4 How are new mountains made?

Extrusive Igneous Rocks

Magma can be forced out onto the Earth's surface through volcanoes to form lava. It cools very quickly and turns into extrusive igneous rock with small crystals.

Movements inside the Earth can push rock from inside the crust up to the surface to make new mountains.

Intrusive Igneous Rocks

Some of the magma cools down inside the Earth's crust. It cools slowly and turns into intrusive igneous rock with big crystals.

Metamorphic Rocks

Metamorphic rocks can be pushed deep into the mantle where they melt back into magma.

1 What is magma?

2 Copy and complete the following sentences.

  a) Magma which cools down on the Earth's crust forms _____ igneous rock.

  b) Magma which cools quickly forms igneous rock with _____ crystals.

Magma

Magma is hot liquid rock.

The rock at the surface of the Earth gets worn away into tiny fragments by weathering and erosion.

? 5 What does weathering do to rock?

6 How are fragments of rock carried away from mountains?

The tiny fragments of rock are carried away from the mountains by gravity, wind, running water or ice.

Tiny fragments of rock sink to the bottom of rivers or the sea and are deposited as sediment. The sediment gets buried under more and more material. The sediment at the bottom is compressed and cemented back into solid rock called sedimentary rock.

Sedimentary Rocks

Igneous and sedimentary rocks can be buried. The huge pressures and very high temperatures underground can change them into metamorphic rocks.

7 How is sediment turned into sedimentary rock?

? 8 What can huge pressures and very high temperatures do to igneous and sedimentary rock?

9 What happens to metamorphic rock when it is pushed deep into the crust?

10 Imagine you are a particle of rock in the magma. Explain how you could end up as a particle in a sedimentary rock.

Summary

Rocks are continually being changed from one form to another. _____ and heat can change sedimentary and igneous rocks into _____ rocks. Metamorphic, sedimentary and _____ rocks can all be pushed down into the _____ where very high temperatures can _____ them into liquid rock called magma. If _____ cools down on the Earth's surface or inside the crust then it forms new igneous rock. If rocks are pushed up to the surface instead they can be worn away by weathering and _____ forming tiny particles called sediment. This _____ can be turned into new sedimentary rock.

erosion   igneous   magma   mantle   melt
metamorphic   pressure   sediment

Tectonic plates

What are tectonic plates?

You probably know that the Earth spins round once a day and travels around the Sun, but did you know that the crust which you are standing on is moving too?

The outer part of the Earth is called the **lithosphere**. The lithosphere is the crust and the outer part of the mantle. It is cracked into a number of big pieces. Each piece of lithosphere is called a **tectonic plate**. These plates are moving very slowly all the time. The map shows the major plates in the Earth's crust. The arrows show which way each plate is moving.

?
1 What are tectonic plates?

2 Which tectonic plate is the UK on?

3 Write down the names of two plates which are moving away from each other.

4 Write down the names of two plates which are moving towards each other.

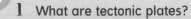

North American Plate

Eurasian Plate

Pacific Plate

Caribbean Plate

Arabian Plate

Philippine Plate

Pacific Plate

African Plate

Nazca Plate

South American Plate

Indo-Australian Plate

Antarctic Plate

A

Tectonic plates move a few centimetres each year. This is about as fast as your fingernails grow

Convection currents

Tectonic plates move very slowly. The huge slabs of lithosphere are moved by very powerful **convection currents** in the mantle. Convection currents are caused by heat, released by the decay (natural beakdown) of radioactive elements inside the Earth.

As the rock in the mantle heats up it becomes less dense than the surrounding rock and rises. It then spreads sideways until it becomes cooler and denser, then it sinks again. As it moves sideways, it carries the tectonic plates with it. Diagram B shows how the convection currents move the plates.

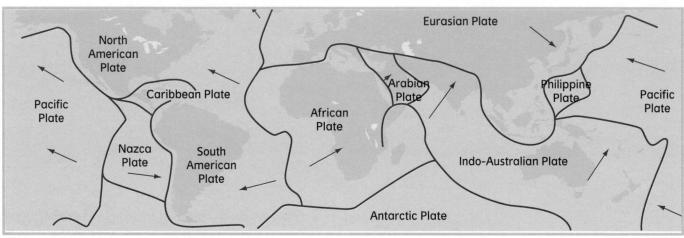

mantle

plate

plate

convection currents

convection currents

B

?
5 On average, how far do tectonic plates move in a year?

6 What causes convection currents?

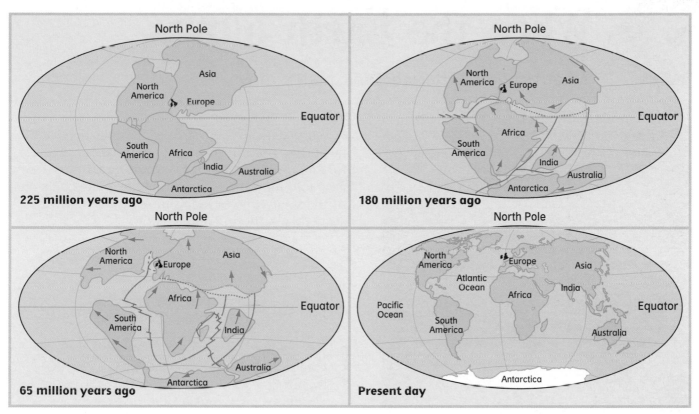

225 million years ago

180 million years ago

65 million years ago

Present day

The moving continents

Scientists have evidence showing that, at one stage in the past, nearly all the continents were joined together. As the plates of lithosphere moved the continents were pulled apart. Since then, the plates have continued to move, taking the continents with them. This is why the Earth looks like it does today. Diagram C shows how the continents have moved.

7 Find the UK on each map in diagram C. Copy and complete the following sentences.

a) As the continents have moved, the UK has got nearer to the _____ (North/South) pole.

b) The UK will have got _____ (warmer/ colder) as it has moved away from the Equator towards the North Pole.

8 Look at map A. Is the UK moving nearer to America or further away?

9 Why do scientists think that the Earth will continue to change?

10 How do convection currents move tectonic plates?

Summary

The Earth's _____ is divided up into pieces called tectonic _____. Each plate is moving very _____ all the time. The force moving the plates comes from _____ currents in the _____. The heat which powers the convection currents comes from _____ decay of elements inside the Earth. All the continents were once joined together. As the plates moved the _____ moved apart.

continents   convection   lithosphere
mantle   plates   radioactive   slowly

When the Earth moves

What happens when tectonic plates move?

When plates move towards each other they can be pushed up, forming huge mountains called fold mountains. Mount Everest is a fold mountain. New fold mountains are formed by plate collisions. At the same time old mountains are being worn away by weathering and erosion.

? 1 How are fold mountains made?

2 Give an example of a fold mountain.

! When scientists measured Everest in 1999, they found it had grown by 2 metres since the last time it was measured in 1954.

A *This is a fold mountain.*

The rock in the crust is put under huge pressures when plates move towards each other. This can change the rock into metamorphic rock (see page 136). Many mountain chains are made of metamorphic rock.

? 3 Why are many fold mountains made of metamorphic rock?

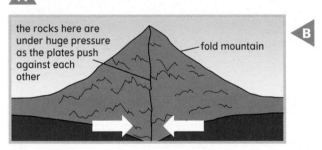

the rocks here are under huge pressure as the plates push against each other

fold mountain

B

When plates move apart, magma from the mantle below rises between them. The magma cools and turns into solid igneous rock. This rock forms a new part of the plate.

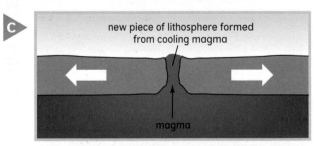

new piece of lithosphere formed from cooling magma

magma

C

? 4 What happens when plates move apart?

Earthquakes and volcanoes

Map D shows the major earthquake zones in the world. It also shows where there are active volcanoes.

? 5 Look at map A on page 140 which showed the plate boundaries. Now look at map D again. What do you notice?

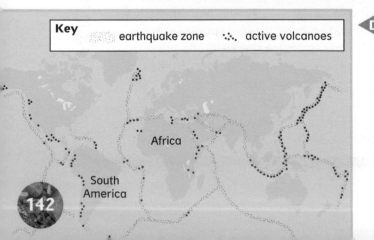

D

Key   earthquake zone   active volcanoes

Africa

South America

Earthquakes happen because plates are moving. As a plate moves past or under another plate it can stick. When enough pressure builds up, the rock breaks and the plate jerks forwards. This makes shock waves that cause an earthquake.

6 Look at map A on page 140. Which two plates meet along the west coast of the USA?

Earthquakes can kill many people if buildings collapse. Scientists can now tell builders where earthquakes are likely to happen. Modern buildings built near plate boundaries now have special foundations which can stop them moving when the earth shakes.

7 Why do buildings near plate boundaries now have special foundations?

Scientists try to predict when earthquakes will happen. They try to work out when the rock will break. This is difficult as it is hard to tell how strong the rock is and how much pressure has built up.

Scientists also try and work out when volcanoes are going to erupt. Just before an eruption scientists can detect movement of magma inside the volcano. Sometimes it pushes the side of the volcano out into a bulge, and this can be measured.

! Over two million people live close to Mount Vesuvius in Italy. Mount Vesuvius has not erupted since 1944, but it could erupt soon!

8 Look at photographs E and F. Why do you think scientists try to predict earthquakes and eruptions?

9 Why is it difficult to predict earthquakes?

10 How can scientists predict when a volcano is about to erupt?

11 Why are some mountains getting higher and some getting lower?

 Collapsed building after an earthquake.

Lava flowing into a town. **F**

Summary

The _____ which make up the Earth's lithosphere are moving all the time. Some plates are moving towards each other while others are moving _____. Some plates are _____ past each other. When plates _____ with each other the crust can be forced up to make _____ mountains. When plates move apart _____ from the mantle rises up between them. When it cools down it forms solid _____ rock. _____ can be caused when plates slide past each other.

apart   collide   earthquakes   fold
igneous   magma   plates   sliding

Alfred Wegener versus the rest

How did scientists find out about plate tectonics?

Have you ever noticed that when fruit like tomatoes and apples get older their skins start to go wrinkly? As the fruit gets older it dries up and the inside shrivels and gets smaller. This makes the skin too big so it wrinkles up.

At one time scientists thought that all the features of the Earth's crust were made in a similar way. They thought that as the Earth cooled down it shrank. If the Earth shrank, the crust would wrinkle and crumple up to form features such as mountain ranges and sea beds. This would mean that all the mountain ranges and sea beds were made at the same time and would all be the same age.

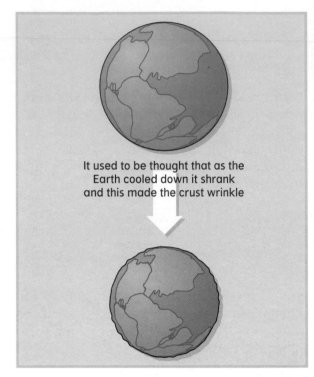

It used to be thought that as the Earth cooled down it shrank and this made the crust wrinkle

A

1 How did scientists think that mountain ranges had been made before they knew about tectonic plates?

Alfred Wegener was a young German scientist who first suggested the idea that the continents were moving around. He thought that all the continents were once joined together as one big land mass which he called Pangaea. He thought that this big land mass then split up to make the different continents which have been moving ever since. When Wegener first suggested this other scientists thought he was mad.

2 Who first suggested that continents were moving around?

3 What was Pangaea?

4 What did Wegener think happened to Pangaea?

 Pangaea.

 Alfred Wegener (1880–1930).

In 1912, Wegener wrote his first book called *The Origin of the Continents and Oceans*. Wegener pointed out that many of the continents had shapes which seemed to fit together. He argued that at one time they must have all fitted together like a jigsaw. Scientists would not believe his idea as they could not understand how big continents could move around.

5 What was the name of Wegener's book?

6 Why did other scientists not believe Wegener's ideas?

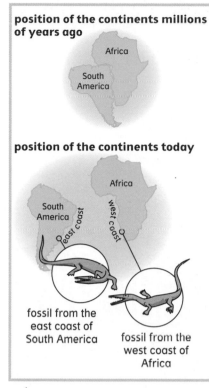

position of the continents millions of years ago

Africa

South America

position of the continents today

Africa

South America

east coast

west coast

fossil from the east coast of South America

fossil from the west coast of Africa

 D

In 1928 a South African scientist called Alex du Toit supported Wegener's idea. He looked at the west coast of Africa and the east coast of South America. He noticed three things.

- The shapes of the two coasts seem to fit together a bit like a jigsaw.
- Both coastlines are made from the same types of rock.
- Both coasts contain fossils of the same kinds of animals and plants.

7 Why did Alex du Toit think that the west coast of Africa and the east coast of South America were once joined up?

In the years that followed more evidence was found that supported Wegener's ideas. Underwater ridges were found. These were like mountain ranges on the sea bed. Many of the ridges had volcanoes. Scientists noticed that the main ridges were half way between continents that Wegener had said were once joined up. They thought that the ridges were the places where the continents had split apart millions of years before.

8 Look at pages 140 and 141. Which two continents are separated by the mid-Atlantic ridge?

Scientists then found out that the crust under the sea was younger than the crust which made up the continents. They realised that as the continents moved apart new sea bed was made. This was why it was younger. Wegener's ideas were finally accepted when scientists realised that convection currents in the mantle would be strong enough to move the huge continents.

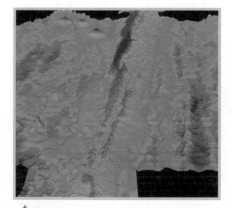

E *Part of the mid-Atlantic ridge.*

9 Which is younger, the crust under the Atlantic Ocean or the crust which makes the land on either side?

10 Why did the discovery that the crust under the sea was younger show that the shrinking earth idea was wrong?

Summary

Scientists used to think that the mountain ranges and sea beds in the Earth's _____ were made when the Earth _____ down and shrank. We now know that the _____ were once joined together in one land mass. This land mass split up into the different continents which then _____ apart from each other. Evidence to support this comes from looking at the shapes of _____ and the rocks and _____ that are found there.

coastlines   continents   cooled   crust   fossils   moved

Home sweet home

How do we make use of Earth materials for building?

Not only is the planet Earth our home, but it also provides us with all the materials we need to build our houses.

Roof slates come from a quarry

Mortar is used to stick stone or bricks together. Mortar is made from sand and cement

Blocks of limestone or sandstone can be used instead of bricks

Concrete path is made from cement, sand and crushed rock

Roof tiles are made from clay

Plastic guttering is made from oil

Bricks are made from clay

Drivers Sold 359694

Window glass is made from sand, with some limestone and sodium carbonate.

A

 1 What are the following things made from?

a) glass
b) mortar
c) concrete
d) bricks and tiles
e) plastics.

Limestone is a very important material for the building trade. It is used to make cement and in making glass. It can also be cut into blocks and used instead of bricks to build the walls of houses. Rocks like limestone are dug out of the ground in **quarries**. Explosives are usually used to blast the rock apart. The pieces of rock can then be cut up into blocks.

B *Limestone is dug out of a quarry.*

 2 a) What is a quarry?
b) What is mortar used for?

3 How is the rock dug out of the ground?

Cement is made by heating powdered limestone and powdered clay in a **rotary kiln**. A kiln is a very big oven and it rotates to mix the limestone and clay. The heat makes the limstone and clay react with each other to make cement

4 What is a kiln?

5 Why is a rotary kiln used?

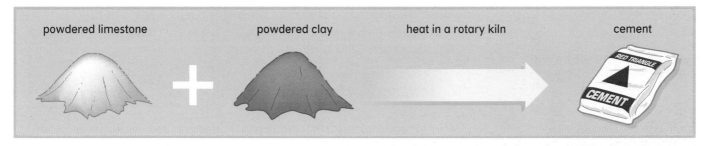

powdered limestone **+** powdered clay → heat in a rotary kiln → cement

Concrete is made by mixing the cement with water, sand and crushed rock. A chemical reaction takes place and the mixture sets into a hard material a bit like stone.

Glass is made by heating pure sand with limestone and soda (soda is another name for sodium carbonate). At first the mixture melts. When it cools down it makes a transparent solid. If something is transparent it means you can see through it.

D

6 Look at diagram C. Use the information above to draw similar diagrams for making concrete and making glass.

P How would you investigate what the best recipe for concrete is?

E

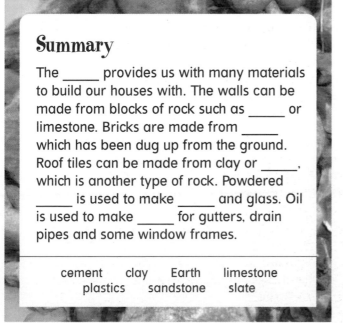

Summary

The _____ provides us with many materials to build our houses with. The walls can be made from blocks of rock such as _____ or limestone. Bricks are made from _____ which has been dug up from the ground. Roof tiles can be made from clay or _____, which is another type of rock. Powdered _____ is used to make _____ and glass. Oil is used to make _____ for gutters, drain pipes and some window frames.

cement	clay	Earth	limestone
plastics	sandstone	slate	

7 a) Name three kinds of rock which are used in the building trade.

b) Are these three rocks igneous, sedimentary or metamorphic rocks? (Hint: you may need to look back at pages 128, 130 and 136.)

Fossil fuels

Where do fossil fuels come from?

A swamp millions of years ago. **A**

Coal, oil and natural gas are all **fossil fuels**. They are called fossil fuels because they were made from the remains of living things which died millions of years ago.

? **1** Write down the names of three fossil fuels.

Coal is made from the remains of plants. Many millions of years ago parts of the Earth were covered by huge forests and swamps. When the trees died they sank to the bottom of the swamps. As time went by, the layers of trees became deeper and deeper as they were covered by more trees which had grown and then died. Later many swamps were covered by rivers which carried **sediment**. The sediment was **deposited** on top of the layers of dead trees.

Over the millions of years which followed more and more sediment was deposited on top of the trees. As there was so little oxygen under all the sediment bacteria couldn't grow and feed on the dead trees. This meant that the trees did not rot away. Instead they were crushed and compressed and eventually turned into coal.

? **2** What was coal made from?

3 What happened to the trees when they died?

4 How were the trees buried and crushed?

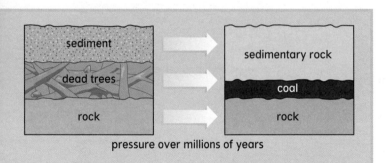

B *Coal formation.*

Oil and gas formation. **C**

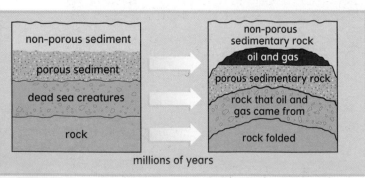

Oil and **natural gas** were made in a similar way to coal although they were made from the remains of tiny plants and animals in the sea. When they died they sank to the bottom of the sea where they were covered by sediment too. As there was no oxygen under the sediment their remains did not decay either.

Over millions of years the sediment turned into sedimentary rock. The pressure from the rock above and the heat from the Earth below slowly released oil and natural gas from the trapped remains.

? **5** What are oil and natural gas made from?

6 Why did the dead trees and animals that formed fossil fuels not decay when they were covered by sediment?

Coal, oil and natural gas are all used as fuels to make electricity at power stations. Oil has many other uses as well. Oil is a mixture of many important chemicals.

 Scientists think we may run out of oil by the year 2030.

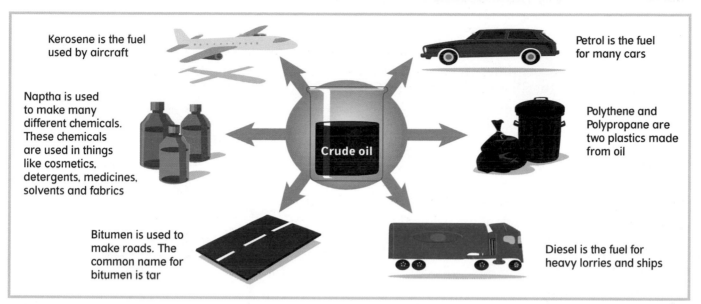

Kerosene is the fuel used by aircraft

Naptha is used to make many different chemicals. These chemicals are used in things like cosmetics, detergents, medicines, solvents and fabrics

Crude oil

Petrol is the fuel for many cars

Polythene and Polypropane are two plastics made from oil

Bitumen is used to make roads. The common name for bitumen is tar

Diesel is the fuel for heavy lorries and ships

When oil is extracted from the ground it is called **crude oil**. It is a thick black sticky mixture of chemicals. The chemicals in any mixture are not chemically joined up to each other. The crude oil is sent to a **refinery** where the mixture of useful chemicals is separated out so that we can use them. The chemicals in oil are called **hydrocarbons**. This means they contain the elements **hydrogen** and **carbon**.

7 What are the following chemicals used for?

a) petrol   **c)** diesel
b) kerosene   **d)** bitumen.

8 Write down the names of two plastics which are made from oil.

9 What happens at an oil refinery?

10 What are hydrocarbons?

11 Are the different chemicals in a mixture chemically joined together or not?

12 Why are coal, oil and natural gas normally found in layers of sedimentary rock?

13 a) Explain how the energy locked in coal comes originally from the sun. (Hint: think how trees and plants make their food.)

b) Explain how the energy stored in oil and natural gas comes originally from the sun. (Hint: small sea creatures eat tiny plants.)

Summary

Coal, _____ and natural gas are all called _____ fuels because they were made from living things which died millions of years ago. Coal was made from layers of dead _____ which were crushed and compressed. Oil and natural _____ were made from the remains of plants and sea animals. Oil is a _____ of many useful chemicals. The chemicals in oil are called _____ because they contain the elements hydrogen and _____.

carbon	fossil	gas	hydrocarbons
	mixture	oil	trees

Fractional distillation

How do we separate the different chemicals in oil?

Crude oil is a thick black sludge that comes out of the ground. It contains many different chemicals which are useful to us, but they have to be separated from one another before we can use them. This is done by heating the crude oil.

All liquids have a **boiling point**. The boiling point is the temperature at which the liquid boils. The boiling point of water is 100 °C. At its boiling point, a liquid will turn into a gas as fast as it can. Another name for a gas is a **vapour**. If the gas cools back down below the boiling point it will **condense** and turn back into a liquid.

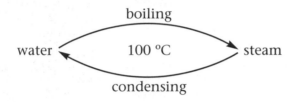

water $\xrightarrow{\text{boiling}}$ steam
$\xleftarrow{\text{condensing}}$
100 °C

1 Look at picture A.

   a) What happens to water when it boils?
   b) What temperature does water boil at?
   c) What happens to steam when it condenses?

A When the steam from the kettle reaches the cold glass in the window it will condense and form drops of water.

All the different chemicals in oil have their own boiling points. Petrol boils at 70 °C. Below 70 °C petrol is a liquid but above 70 °C it will be a vapour. Kerosene boils at 180 °C. Below 180 °C kerosene is a liquid but above 180 °C it will be a vapour.

2 What is a vapour?

3 **a)** Will diesel be a liquid or a vapour at 280 °C?
   b) Will diesel be a liquid or a vapour at 230 °C?
   c) What happens when diesel vapour cools down below 260 °C?

Distillation is when you turn a liquid into a gas and then back into a liquid. **Fractional distillation** can be used to separate liquids with different boiling points. All the different chemicals in oil are separated by fractional distillation. Each chemical which is separated out is called a **fraction**. This happens in a oil refinery.

Chemical	Boiling point
Petrol	70 °C
Naphtha	140 °C
Kerosene	180 °C
Diesel	260 °C
Bitumen	360 °C

 B

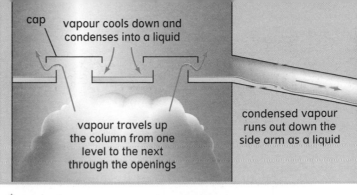

cap — vapour cools down and condenses into a liquid

vapour travels up the column from one level to the next through the openings

condensed vapour runs out down the side arm as a liquid

The different chemicals in oil are separated in a **fractionating column**. The column has different levels. There are openings with caps between levels.

4 Look at diagram C

a) How can vapours move up from one level to the next?

b) If a vapour condenses back into a liquid where does it go?

5 Look at diagram D.

a) What happens to the temperature of the column as you move up?

b) Which fraction is separated out at level 4?

c) Which fraction separates out without condensing into a liquid?

d) How many fractions are separated out altogether?

A fractionating column.

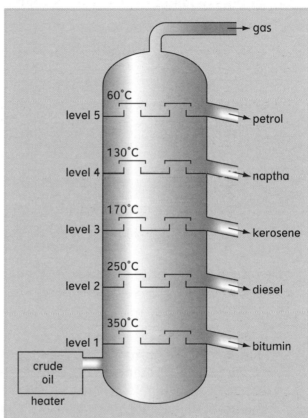

gas

60°C — level 5 — petrol

130°C — level 4 — naptha

170°C — level 3 — kerosene

250°C — level 2 — diesel

350°C — level 1 — bitumin

crude oil

heater

P How would you separate ethanol and water by fractional distillation?

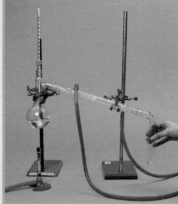

E

6 a) What happens to the petrol in the crude oil mixture when the temperature in the heater reaches 70 °C?

b) Where does the petrol vapour go?

c) What happens to the petrol vapour when it reaches level 5?

d) Why does the vapour not condense until it reaches level 5?

7 a) What happens to the diesel in the crude oil mixture when the temperature in the heater reaches 260 °C?

b) Where does the diesel vapour go?

c) What happens to the diesel vapour when it reaches level 2?

d) Why does the vapour not condense at level 1?

Summary

The different fractions in crude oil are separated out in a _____ column by fractional _____. The different fractions have different _____ points. As each fraction reaches its boiling point it turns into a _____ and travels up the column. When the vapour cools down it condenses back into a _____ and drains out.

boiling   distillation   fractionating
liquid   vapour

Burning fuels

What happens when we burn fuels?

When an element burns it joins up with oxygen and forms an oxide. Fuels like coal, oil and natural gas contain the elements carbon and hydrogen. Coal and oil also contain some sulphur. When the fuels burn these elements combine with oxygen to form their oxides.

carbon + oxygen ⟶ carbon dioxide

hydrogen + oxygen ⟶ hydrogen oxide
(hydrogen oxide is better known as water!)

sulphur + oxygen ⟶ sulphur dioxide

1 What three elements are found in fuels?

2 What oxides are made when fuels burn?

3 What is hydrogen oxide better known as?

When fuels burn, the oxides that are made are nearly always gases. Carbon dioxide, sulphur dioxide and water vapour are all gases which are made when coal, oil and natural gas are burnt

Each year we burn huge amounts of fossil fuels. Coal, oil and natural gas are burnt at power stations to make electricity. Petrol, diesel and kerosene are all burnt to power our cars, lorries, ships and aircraft. So much fossil fuel is burnt that we add more than 25 000 000 000 tonnes of carbon dioxide and sulphur dioxide to our atmosphere each year.

Sulphur dioxide is an acid gas. When it dissolves in the clouds it makes **acid rain**. When acid rain falls it makes soil more acidic. Many plants cannot live or grow in very acid soil. Eventually even trees can be killed and large forests can be destroyed. Acid rain can also make water too acidic for animals to live in. Fish can no longer survive in many lakes and rivers around the world.

4 Which gas causes acid rain?

5 How can acid rain affect plant life?

6 Why do many lakes no longer have any fish living in them?

Scientists say that carbon dioxide is a **greenhouse gas** because it traps the Sun's heat in our atmosphere and makes the Earth warmer. This increase in the Earth's temperature is called **global warming**.

A

B

Scientists predict that the Earth's temperature will continue to rise as we produce more and more carbon dioxide. They are worried that the polar ice caps will start to melt and the sea level around the world will start to rise. This will cause serious flooding in some countries. Changes to the Earth's climate may make other parts of the world too dry to grow crops, and this could lead to food shortages and famine.

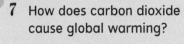

D

Governments from around the world are looking at ways that we can cut down on the amount of fossil fuel that we burn. This will reduce the amount of carbon dioxide and sulphur dioxide that we are making.

E

F

7 How does carbon dioxide cause global warming?

8 Why do scientist think that global warming may lead to serious flooding?

9 Why do sicentists think that global warming may lead to famine?

G

10 Look at the photographs E, F and G. Write down three ways that we can reduce the amount of fossil fuel that we burn.

11 All our electricity used to come from fossil fuelled power stations. Find out other ways that we can now generate electricity.

Summary

When we burn _____ fuels the gases carbon dioxide, _____ _____ and water vapour are made. _____ _____ is a greenhouse gas and is trapping _____ in the Earth's atmosphere. This is causing _____ warming. Sulphur dioxide dissolves in water in the clouds to make _____ rain.

| acid | carbon dioxide | fossil |
| global | heat | sulphur dioxide |

Limestone and acid rain

How can limestone help the environment?

Acid rain can cause many problems for the environment (see page 152). It can make soil too acidic for plants to grow. Some rivers and lakes around the world are too acidic for fish to live in. Scientists now use limestone to treat acid conditions so that living things can survive there again.

A

B

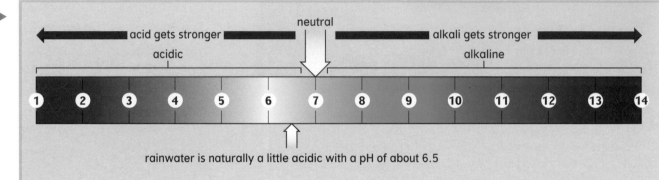

neutral

← acid gets stronger ▬ | alkali gets stronger ▬ →

acidic | alkaline

1 2 3 4 5 6 7 8 9 10 11 12 13 14

↑ rainwater is naturally a little acidic with a pH of about 6.5

The pH scale measures how acidic or alkali something is. The pH scale goes from 1 to 14.

1 Look at diagram B.

   a) What pH is neutral?
   b) Which is a stronger acid, pH 2 or pH 5?
   c) Which is a stronger alkali, pH 8 or pH 13?
   d) What pH is rainwater normally?

If you **neutralise** an acid solution you turn it from being acid to being neutral. You can neutralise an acid by adding a **base** to it. Bases that dissolve in water are called alkalis. An alkali can also cancel out an acid to make something neutral.

2 How can you neutralise an acid?

The main chemical in limestone is **calcium carbonate**. Calcium carbonate is a base. Powdered limestone can be added to acidic rivers and lakes to neutralise the water. This means that fish and other animals can live there again.

! The strongest acid rain ever recorded had a pH of 2. It damaged the paint on cars when it rained.

C *Powdered limestone being added to a lake.*

3 Why is powdered limestone added to acidic rivers and lakes?

Farmers and gardeners often use a base called **slaked lime** to neutralise acidic soil so that they can grow crops or flowers. Slaked lime is made from limestone.

 4 How can farmers and gardeners neutralise their soil if it is too acid?

Slaked lime is made from limestone in two stages

Stage 1

First of all the limestone is heated in a big oven called a kiln. When it is heated the calcium carbonate in the limestone changes into **quicklime** and carbon dioxide. The chemical name for quicklime is **calcium oxide**.

 5 How is limestone changed into quicklime?

This kind of reaction is called **thermal decomposition**. Thermal means heat. Decomposition means to break down into smaller bits. Thermal decomposition just means that a chemical breaks down into smaller chemicals when it is heated.

Stage 2

Water is then added to quicklime. The quicklime reacts with water to make **slaked lime**. The chemical name for slaked lime is **calcium hydroxide**.

 6 How is quicklime changed into slaked lime?

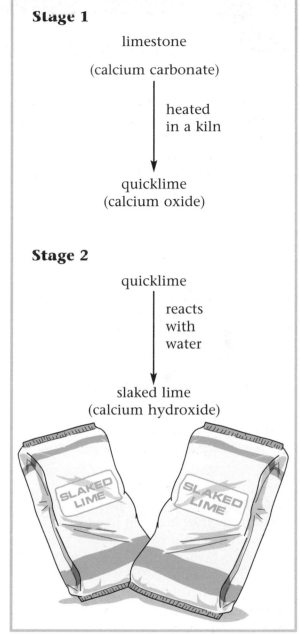

Stage 1

limestone
(calcium carbonate)

heated in a kiln

quicklime
(calcium oxide)

Stage 2

quicklime

reacts with water

slaked lime
(calcium hydroxide)

SLAKED LIME   SLAKED LIME

Summary

Acid rain can make _____ too acidic for plants to grow. It can also make _____ and lakes too acidic for _____ to live in. Acidic rivers or lakes can be _____ by adding powdered _____ . Acidic soil can be neutralised by adding _____ _____ .
Slaked lime is made by heating limestone in a kiln to make _____ . The quicklime then reacts with _____ to make slaked lime.

| fish | limestone | neutralised | rivers |
| slaked lime | soil | quicklime | water |

7 Write down the chemical names for:

  a) limestone
  b) quicklime
  c) slaked lime.

8 What is made in the thermal decomposition of limestone?

9 How would you neutralise soil or water that was too alkaline?

More about oil

Why do the chemicals in oil have different properties and uses?

Oil is a mixture of different chemicals such as petrol, naphtha and diesel. These chemicals are all hydrocarbons. Some of the chemicals in oil are long hydrocarbon molecules which have a lot of carbon atoms joined up together in a chain. Others are smaller molecules with fewer carbon atoms in the chain.

Hydrocarbons with long chains of carbon atoms have high boiling points. Hydrocarbons with shorter chains of carbon atoms have lower boiling points.

A

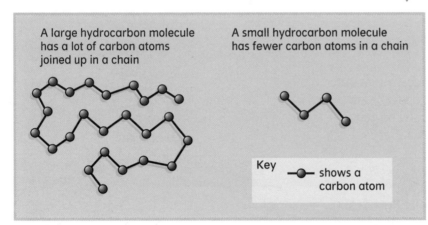

A large hydrocarbon molecule has a lot of carbon atoms joined up in a chain

A small hydrocarbon molecule has fewer carbon atoms in a chain

Key —●— shows a carbon atom

Long chain hydrocarbon molecules form very **viscous** liquids. This means they are a bit like treacle and will not flow very easily. Short chain hydrocarbons are very runny. Short chain hydrocarbon molecules are also very **flammable**. This means that they will burn easily. Long chain hydrocarbon molecules are not very flammable. This means they are not very good as fuels.

 2 What is a viscous liquid?

3 What is a flammable chemical?

Number of carbon atoms	Size of molecule	Boiling point	Thickness (viscosity)	Flammability
8	small	60 °C	runny	easy to light
15	medium	150 °C	quite thick	harder to light
50	large	350 °C	very thick	difficult to light

C

1 Look at table B.

a) Which chemical has the longest chain of carbon atoms?

b) Which chemical has the shortest chain of carbon atoms?

c) Which is the longer chain hydrocarbon, kerosene or diesel?

B

Chemical	Boiling point
Petrol	70 °C
Naphtha	140 °C
Kerosene	180 °C
Diesel	260 °C
Bitumen	360 °C

4 Copy and complete the following sentences using the words 'longer' or 'shorter.'

a) _____ chain hydrocarbons are runny liquids.

b) The _____ the hydrocarbon chain the harder it is to set on fire.

c) The _____ the hydrocarbon chain the lower the boiling point.

Cracking

After all the useful chemicals have been separated out of the crude oil, the oil companies are left with a mixture of long chain hydrocarbon molecules which they do not really need. These long chain hydrocarbon molecules do not have many uses. Scientists have found a way to break up these long chain hydrocarbon molecules into smaller chain hydrocarbon molecules which they can use.

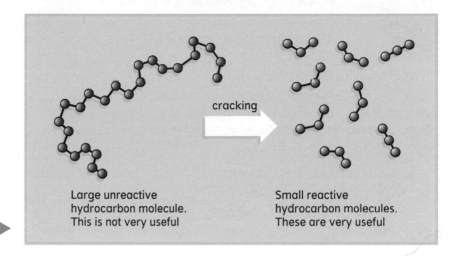

Large unreactive hydrocarbon molecule. This is not very useful

cracking

Small reactive hydrocarbon molecules. These are very useful

D

First of all the large hydrocarbon molecules are heated. This changes them from a liquid into a gas or **vapour**. The vapour is then passsed over a hot catalyst. A **catalyst** is a chemical which speeds up a reaction. In this case, the catalyst speeds up the breakdown of large hydrocarbon molecules into smaller ones.

Breaking down large hydrocarbon molecules into smaller, more useful ones is called **cracking**.

This reaction is another example of **thermal decomposition** where a chemical breaks down into smaller chemicals when it is heated (see page 155).

?

5 Which are more useful, short or long chain hydrocarbon molecules?

6 What is cracking?

7 What is a catalyst?

8 How are large hydrocarbons cracked?

9 Why do you think bitumen makes a better road surface than a fuel? (Hint: think of how runny it is and how flammable it is.)

10 Why do you think petrol makes a better fuel than a road surface?

11 Why do you think oil companies developed the process of cracking?

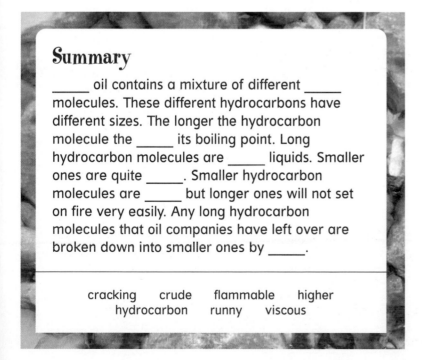

Summary

_____ oil contains a mixture of different _____ molecules. These different hydrocarbons have different sizes. The longer the hydrocarbon molecule the _____ its boiling point. Long hydrocarbon molecules are _____ liquids. Smaller ones are quite _____. Smaller hydrocarbon molecules are _____ but longer ones will not set on fire very easily. Any long hydrocarbon molecules that oil companies have left over are broken down into smaller ones by _____.

cracking   crude   flammable   higher
hydrocarbon   runny   viscous

Polymers

What are polymers?

Scientists **crack** long hydrocarbon molecules into smaller molecules. They can then use them to make new materials such as plastics.

1 What happens to large molecules when they are cracked?

2 What are plastics made from?

A *Plastic shopping bags are made from polythene.*

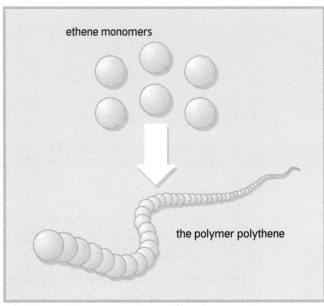

ethene monomers

the polymer polythene

B

One smaller molecule produced by cracking is called **ethene**. Ethene molecules are very reactive. This means they join up very easily with other molecules. When ethene molecules are heated they join together to make a giant molecule called **polyethene**. Poly means a lot, so polyethene means a lot of ethene molecules joined up together.

You will know polyethene better as **polythene** which is the plastic used to make carrier bags, bin liners and plastic sheeting.

Giant molecules like polythene which are made from smaller reactive molecules are called **polymers**. The small reactive molecules which join up to make the polymers are called **monomers**.

3 Write down an example of a monomer.

4 Write down an example of a polymer.

5 How is polythene made?

6 Copy and complete table C by matching up each polymer with the monomer it was made from. The monomers you can choose from are:

propene   ester   styrene

C

Polymer	Monomer
polystyrene	
polypropene	
polyester	

Polypropene, polyester and polystyrene are all useful polymers.

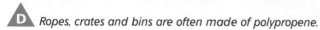
D Ropes, crates and bins are often made of polypropene.

E Packaging, disposable cups and food containers are often made of polystyrene.

F Many clothes contain some polyester.

7 Name some things which are made from:

a) polypropene **b)** polyester **c)** polystyrene.

Plastics are very useful materials but there is one problem with them. Plastics are not **biodegradable**. This means that micro-organisms cannot break them down so they will not rot. This can lead to problems getting rid of them when they are finished with.

8 Explain why getting rid of plastic waste can be a problem.

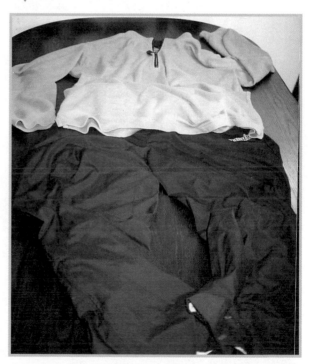

Summary

_____ are giant molecules. They are made of lots of _____ reactive molecules called _____ which are joined together. Plastics such as polythene and polypropene are polymers. Polythene is made from the monomer _____. Polypropene is made from the monomer _____.

ethene   monomers   polymers
propene   small

9 What properties does polystyrene have which makes it a good packaging material?

10 What properties does polythene have which makes it good for making carrier bags?

D18 The early atmosphere

Where did the atmosphere come from?

The atmosphere formed during the first billion years after the Earth was made. At this time the surface of the Earth was covered in volcanoes. When they erupted they gave out gases. These gases formed the atmosphere.

1 a) What were the two main gases given out by volcanoes?
  b) Which two other gases were in the atmosphere?

2 How were the seas and oceans made?

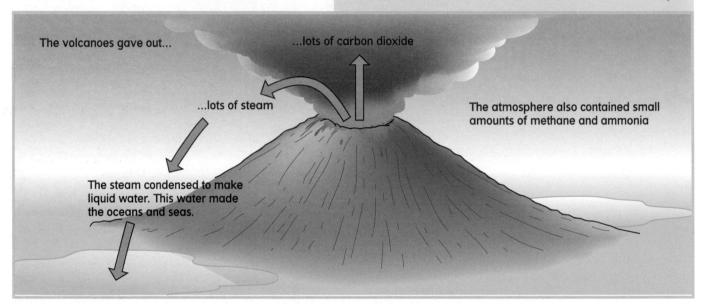

The volcanoes gave out...

...lots of carbon dioxide

...lots of steam

The atmosphere also contained small amounts of methane and ammonia

The steam condensed to make liquid water. This water made the oceans and seas.

Some other planets in our solar system have atmospheres too. When our atmosphere was first made it was nearly all carbon dioxide with very little or no oxygen at all. Scientists think that the atmospheres on Venus and Mars are like that today.

The first living things to appear on Earth were simple bacteria. As there was no oxygen in the atmosphere, these bacteria could not have used aerobic respiration to get energy. Living things which can live without oxygen are called **anaerobes**.

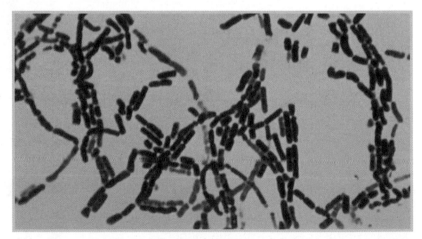

Anaerobic bacteria.

3 a) What were the first living things to appear on Earth?
  b) What are anaerobes?

! The largest volcano in our solar system is on Mars. It is 500 km wide and 25 km high. This is 100 times bigger than any volcano on Earth.

The ozone layer in the top of the atmosphere keeps out harmful ultraviolet rays from the sun. Ozone is made from oxygen. When the atmosphere was first made it had no ozone layer as there was no oxygen. This meant that the first living things had to live under water where they were protected from the harmful ultraviolet rays.

4 Why do we need the ozone layer?

5 Why was there no ozone layer when the atmosphere was first made?

The first green organisms that appeared on Earth were small one-celled algae which lived in the sea. They made their food by photosynthesis, like plants.

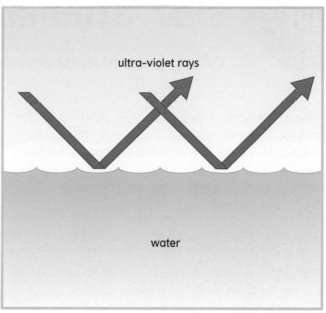

Ultraviolet rays are reflected from the surface of the water.

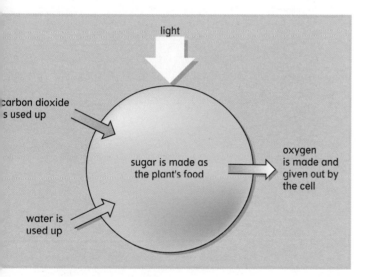

An algal cell.

6 a) What gas do algae and plants use up in photosynthesis?

b) What gas do algae and plants make in photosynthesis?

7 Copy and complete the following sentences using the words 'increased' or 'decreased'.

a) When the algae photosynthesised, the amount of carbon dioxide in the atmosphere would have _____ as it was used up.

b) The amount of oxygen in the atmosphere would have _____ due to the algae photosynthesising.

Summary

When the Earth was made the surface was covered in _____. When the volcanoes _____ they gave out gases which made the _____. The main gases were carbon dioxide and _____ although there was some _____ and ammonia too. When the steam cooled and _____ it formed the oceans and the seas. The atmosphere did not contain _____. The atmosphere started to change when green _____ photosynthesised. They used _____ _____ and made oxygen.

algae   atmosphere   carbon dioxide
condensed   erupted   methane
oxygen   steam   volcanoes

8 How did the Earth's atmosphere start to change over the first billion years?

9 Scientists have found out that there is now a hole in the ozone layer.
Why are they worried by this?

161

Our atmosphere now

What is our atmosphere like today?

When our atmosphere was first made it had no oxygen. The first green organisms that appeared on Earth were small one-celled algae which lived in the sea. When they photosynthesised they started to add oxygen to the atmosphere. The oxygen had three effects.

- Some bacteria that had first lived on the Earth were poisoned by the oxygen and died out.
- Methane and ammonia in the air reacted with the oxygen to make carbon dioxide and nitrogen and water.
- Ozone was made from the oxygen. This made a layer in the top of the atmosphere which kept out harmful ultraviolet rays. Now that the Earth had an ozone layer living things did not have to stay under water. Plants and animals could now live on land.

1 Why did some bacteria die out when oxygen was added to the air?

2 What happened to the methane and ammonia in the atmosphere?

3 Why could animals and plants now live on land?

Over the millions of years that followed new animals and plants appeared. The land became covered in thick forests. Slowly the atmosphere changed. The first amphibians evolved and began to live on land 380 million years ago. The first mammals appeared about 200 million years ago.

4 When did the first mammals appear on Earth?

The atmosphere today is quite different from the Earth's first atmosphere. Pie chart B shows what is in the atmosphere today.

5 a) Which two gases make up most of the atmosphere today?
   b) Name two other gases in the atmosphere.

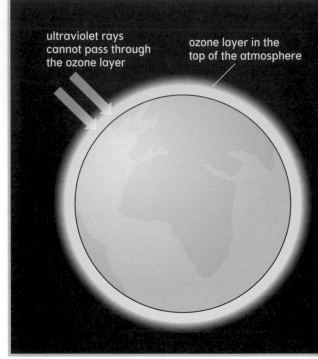

ultraviolet rays cannot pass through the ozone layer

ozone layer in the top of the atmosphere

A

B

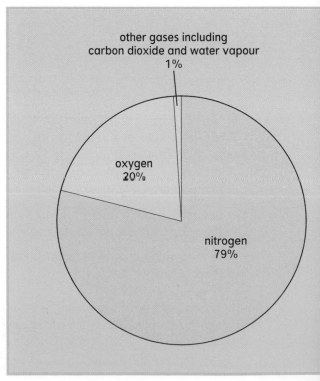

other gases including carbon dioxide and water vapour
1%

oxygen
20%

nitrogen
79%

When the atmosphere was first made it was nearly all carbon dioxide. Now the amount of carbon dioxide is less than 1 %. Carbon dioxide has carbon in it, so where has all this carbon gone?

A lot of carbon dioxide dissolved in the sea to make chemicals called carbonates. These carbonates were deposited as sediment on the sea bed. Over millions of years the carbonates became sedimentary rocks. A lot of carbon from the early atmosphere is now found in carbonates in rocks in the Earth's crust.

 6 What chemicals were made when the carbon dioxide dissolved in the sea?

7 What happened to the carbonates that were made?

Some sea animals used the carbonates in the water to make calcium carbonate. They used the calcium carbonate to make their shells. When the animals died their shells were deposited as sediment. Over millions of years the shells became limestone. A lot of carbon is now found in calcium carbonate in limestone.

 8 Name a sedimentary rock made from calcium carbonate?

A lot of carbon is also found in the fossil fuels coal, oil and natural gas. Coal was formed from trees which died millions of years ago. When the trees were alive they took in carbon dioxide. They used the carbon to make new cells as they grew. When the trees were crushed the carbon became part of the coal. Oil and natural gas were formed from the remains of tiny plants and animals in the sea. The carbon that was in them became part of the oil and gas.

9 How did carbon get trapped in coal?

 C *Seashells are made of calcium carbonate.*

Summary

As the atmosphere changed the amount of oxygen _____ and the amount of carbon dioxide _____. Now 79 % of the atmosphere is _____, 20 % is _____ and less than 1 % is _____ _____. A lot of the carbon that was in the atmosphere is now locked up in _____ rocks like _____ or in fossil fuels such as coal, _____ and natural gas.

carbon dioxide	decreased
increased limestone	nitrogen
oil oxygen	sedimentary

10 Use the information in pie chart B to draw a simple bar chart to show the gases in the atmosphere.

11 How do you think carbon in fossil fuels is released back into the air?

12 Find out four things that limestone is used for.

 D *Fossil fuels contain a lot of carbon*

Further questions

1   This drawing shows the three main layers of the Earth.

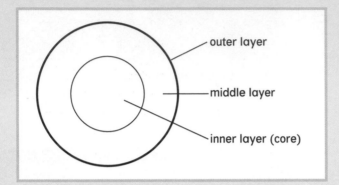

Copy and complete these sentences using words from the box. You may use each word once, more than once, or not at all.

The outer layer is called the _____. It is made from _____. The middle layer is called the _____. The inner layer is called the _____. It is made from _____ and nickel. The inner part of it is _____ and the outer part is _____. (7)

atmosphere	core	crust
liquid	gas	iron	mantle
rock	solid	

2   This diagram shows rocks in a cliff face.

a) i) Which rock is likely to be the oldest?
   ii) Explain your answer. (2)
b) Are rocks B, C and E igneous, sedimentary or metamorphic rocks? (1)

c) What are fossils? (3)
d) This diagram shows some rocks in a nearby hillside.

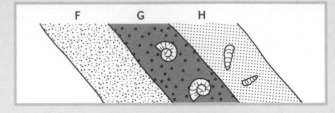

i) Which rocks are the oldest?
ii) Explain your answer. (2)

3   This map shows Africa and South America.

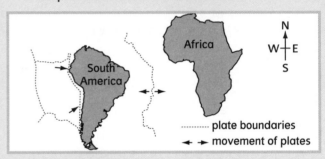

a) Are South America and Africa moving closer to each other or further away? (1)
b) Give three reasons why scientists think that South America and Africa were once joined together. (3)
c) What makes tectonic plates move? (1)

4   Copy and complete these sentences using words from the box. You may use each word once, more than once, or not at all.

_____ is made by heating powdered clay and limestone in a kiln.

_____ is made by heating limestone in a kiln.

_____ _____ is made by adding water to quicklime. (3)

cement	limestone	quicklime
sand	slaked lime	

5 Look at the diagram of a fractionating column.
It is used to separate the hydrocarbons in crude oil.

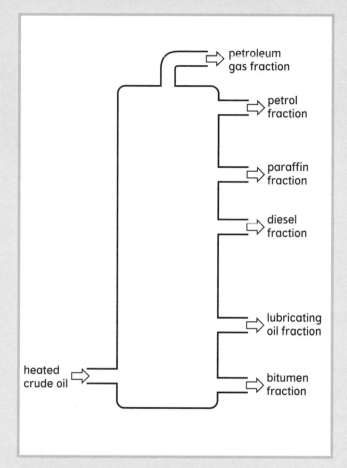

a) Which two elements do hydrocarbons contain? (2)

b) Which part of the fractionating column will be the hottest? (1)

c) Which fraction shown in the diagram will have the lowest boiling point? (1)

d) i) Which fraction shown in the diagram will be the most viscous?

ii) Explain your answer as fully as you can. (3)

6 Most fuels contain carbon and hydrogen.

a) Which other element do some fuels contain? (1)

b) Name three gases that are formed when fossil fuels burn. (3)

c) What is global warming? Explain in as much detail as you can. (3)

d) i) Why is acid rain a problem?

ii) Which gas causes acid rain? (3)

7 Copy and complete these sentences using words from the box. You may use each word once, more than once, or not at all.

Some of the long chain _____ molecules obtained from _____ oil are not very useful. They can be broken up into _____ molecules by _____. These smaller molecules are called _____ and can be joined together to make _____. Lots of ethene monomers can be joined together to make _____. This is a useful plastic which is used to make _____ materials such as bags and bottles. (8)

bigger	cracking	crude
hydrocarbon	monomers	packaging
plastic	polymers	polystyrene
polythene	smaller	

8 These pie charts show the composition of the atmosphere at different times in the history of the Earth.

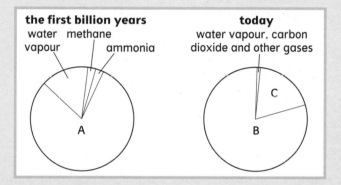

a) What are the gases labelled A, B and C? (3)

b) Where did the gases in the early atmosphere come from? (2)

c) How were the oceans formed? (2)

d) The atmosphere today contains oxygen. Where did the oxygen come from? (2)

e) Fossil fuels and some kinds of rock contain a lot of carbon. Where did this carbon come from? (1)

Conduction

How is heat transferred through solids?

Hot drinks from fast food restaurants are often served in polystyrene cups to keep them warm. Polystyrene contains lots of little air pockets. Heat does not travel well through the trapped air so the cup loses heat slowly.

A

 1 How do polystyrene cups keep drinks warm?

When hot objects lose heat, they cool down. Anything hotter than its surroundings loses heat until it reaches the same temperature as the surroundings. Heat energy always flows from a hot place to a cooler place. Opening a door on a cold day does not let the cold in; it lets the heat out!

When heat flows from place to place we say it **transfers**.

 2 Copy diagrams B and C. Add arrows to each diagram to show which way the heat flows.

B   C

Heat energy stops flowing when the objects reach the same temperature. The hotter objects stop cooling down and the colder objects stop heating up. A cup of coffee cools down until it reaches room temperature. It will never become colder than room temperature unless it is put into an even colder place like a fridge.

D   Heat flows from hotter places to cooler places.

 3 Look at picture D. What happens to the temperature of the air surrounding the coffee cup?

Everything is made of particles. In solids, the particles are close together in fixed positions. The particles cannot move about, but they can vibrate. As solids become hotter, their particles vibrate more. These vibrations pass their heat energy on to nearby particles. This method of heat transfer is called **conduction**, and mainly takes place in solids.

 4 Explain how heat is transferred in solids.

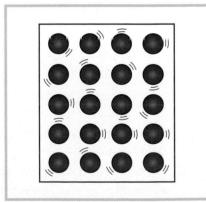

E   In solids, the particles are close together and cannot change places.

Solids that transfer heat well are called **heat conductors**. Metals are good heat conductors. Non-metals and gases are poor heat conductors, and are also called **heat insulators**. Heat insulators transfer heat very slowly.

5 What do we mean by heat insulators?

6 Write down an example of one good heat conductor and one heat insulator.

F *Metals conduct heat well.*

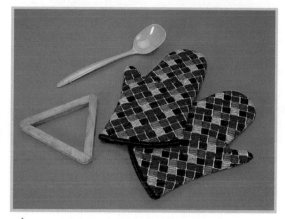

G *Non-metals are good insulators.*

 Houses can lose heat through windows and window frames. Window frames can be made of many different materials.

● Which material is best at preventing heat losses through a window frame?

● How would you find out?

● How would you make sure your experiment is a fair test?

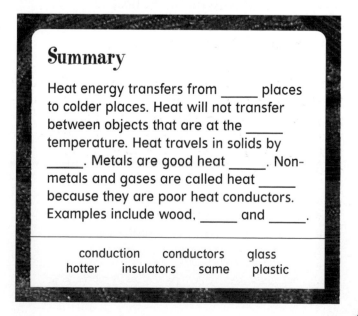

H

! Special silica tiles are such good insulators that they can protect astronauts inside space capsules returning to earth from temperatures on the outside of the capsule of over 1500 °C.

7 a) Why does a hot piece of toast cool down?
   b) After a while, the toast stops cooling down. Why?

8 Do your fingers get hotter stirring hot drinks with a metal spoon or a plastic spoon? Explain your answer.

9 Carpets help to stop heat losses. They are often made of wool, which contains small pockets of trapped air.
   a) Why do wool carpets keep the heat in?
   b) Why are thicker carpets better at stopping heat losses?

Summary

Heat energy transfers from _____ places to colder places. Heat will not transfer between objects that are at the _____ temperature. Heat travels in solids by _____. Metals are good heat _____. Non-metals and gases are called heat _____ because they are poor heat conductors. Examples include wood, _____ and _____.

conduction	conductors	glass
hotter	insulators	same	plastic

Convection

How is heat is transferred through gases and liquids?

Smoke detectors save lives by warning people that their house is on fire. Smoke detectors are always fitted on ceilings because the hot smoke rises and can be detected quickly.

? 1 Why should you crouch down in a smoke filled room?

**! ** Breathing in smoke kills more people in house fires than the flames do.

A

Liquids and gases are called **fluids**. They can flow because they do not have fixed shapes like solids. When the tiny particles in fluids flow, they carry heat energy with them. This is called **convection**. The heat travels quickly with the particles as they move.

? 2 Write down two ways that gases and liquids are different from solids.

3 What do we mean by convection?

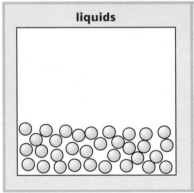

liquids	gases

B In liquids, the particles are close together but can swap places

C In gases, the particles are far apart and can swap places

Convection currents allow the heat to transfer evenly throughout a liquid or gas. The particles of hot liquids and gases rise. The particles of colder liquids and gases sink. This can create a **convection current**.

? 4 Does hot water rise or sink?

Convection can warm things up. You can use convection to transfer heat quickly. The heating element of a kettle is at the bottom. As the water at the base of the kettle heats up, it rises. It is replaced by the cooler water above it, which sinks.

As hot liquid rises, cooler liquid takes its place.
This forms a convection current. **D**

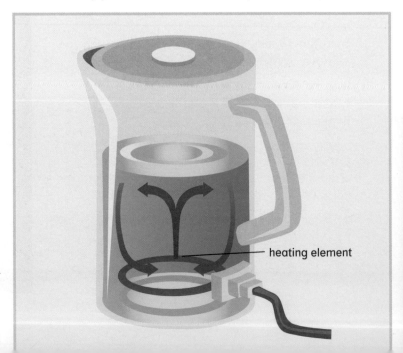

heating element

P How would you design an experiment to decide where the coolest air surrounding a hot object is?

Convection can cool things down. The cooling element for a fridge is in the icebox, which is at the top of the fridge. The cooling element transfers heat out of the fridge.

The air is colder near the icebox, and sinks to cool the lower part of the fridge. The warmer air rises from the bottom of the fridge. This way, the circulating air cools down quickly.

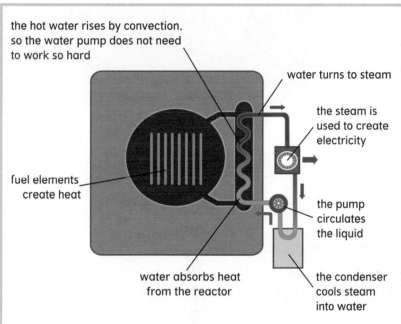

the hot water rises by convection, so the water pump does not need to work so hard

water turns to steam

the steam is used to create electricity

fuel elements create heat

the pump circulates the liquid

water absorbs heat from the reactor

the condenser cools steam into water

5 Why is the fridge icebox at the top of the fridge?

Nuclear reactors change water to steam. Convection helps to heat the water more effectively.

H *A nuclear reactor.*

6 Look at diagram H. How does convection help a nuclear reactor heat water, creating steam?

7 Why is the top part of the oven the hottest part?

8 How does a hat help to keep you warm on a windy day? Choose the correct answer.

   A) It stops the wind blowing away the warm air from around your head.
   B) It looks cheerful.
   C) It stops your hair getting tangled.

9 Explain why a hot air balloon floats in the sky.

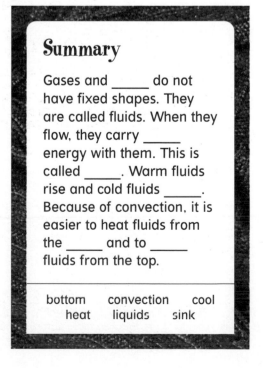

Summary

Gases and _____ do not have fixed shapes. They are called fluids. When they flow, they carry _____ energy with them. This is called _____. Warm fluids rise and cold fluids _____. Because of convection, it is easier to heat fluids from the _____ and to _____ fluids from the top.

bottom	convection	cool
heat	liquids	sink

E3 ▶ Radiation

How is heat transferred through space?

Studying volcanoes is hot work! Scientists working near an eruption wear white or silver suits, which reflect the heat from the volcano. This keeps the scientists cool enough to survive.

 1 Why do white clothes help you keep cool?

Hot objects give out heat. They are said to **radiate** heat. **Heat radiation** is also called **infra-red radiation**. Hotter objects radiate more heat. A big fire at an oil refinery is much hotter than a bonfire. Fire fighters at the oil refinery need more protection from radiated heat and so they wear white suits.

Radiation travels as waves of energy and this is how the Sun's heat reaches the Earth. Radiation can travel through empty space and through transparent things like air or glass. Radiation does not travel using particles.

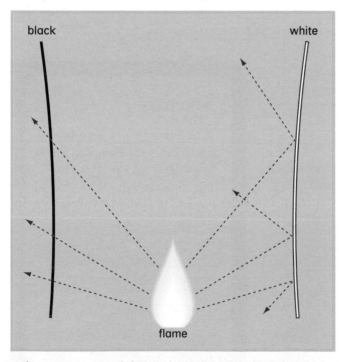

black   white

flame

 B *White surfaces reflect radiated heat so the sheet stays cool to touch. Black surfaces absorb radiated heat which then radiates from the other side.*

A

 2 What is another name for heat radiation?

 3 How does radiation travel?

Reflecting radiation

Some colours are better than others at reflecting heat. White and shiny surfaces reflect heat well. Lorries carrying milk or chocolate are often painted white. This **reflects** the Sun's heat so the milk or chocolate inside stays cool.

Giving out heat radiation

Black surfaces are good at radiating heat. The insides of ovens are a dull black colour to help the food cook more quickly.

Silvery and white surfaces also do not radiate heat well. A person standing near a white car on a hot day feels less heat radiating from its paintwork than from a black car.

 4 Which colours are best at reflecting heat radiation?

Absorbing radiation

Solar panels **absorb** the sun's heat and use it to heat water. They are black because black absorbs heat best.

	Black	Silver
Reflecting heat radiation	worst	best
Giving out heat radiation	best	worst
Absorbing heat radiation	best	worst

5 Which colour is best at absorbing heat radiation?

6 Would astronauts approaching the Sun stay coolest for longer in a white spaceship or in a black spaceship? Explain your answer.

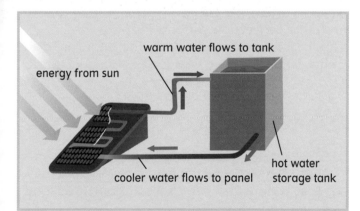

E *This solar panel is designed to heat water. Black surfaces absorb the Sun's heat well.*

Summary

Heat travels as waves through space. This is called ____. Hotter objects radiate ____ heat than cooler objects. ____ surfaces absorb and give out radiated heat well. Shiny surfaces do not absorb heat well: they ____ radiation. Shiny surfaces are used if you want to ____ radiation.

black   more   radiation   reflect   stop

P Most people like their cold drinks to stay cool while they drink them. How would you find out the best colour for a soft drink can?

D

7 Look at diagram G. Which cup radiates most heat and cools down quickest? Explain your answer.

G

8 Wrapping food in aluminium foil stops it cooking too fast. How does this work? Choose the correct answer.

A) By stopping heat conduction.
B) By reflecting the heat radiation.
C) By cooling the food.

9 Some frozen pies are sold in foil dishes painted black on the outside. Why does this help them to cook better?

10 Your house stays cool in summer if the curtains are drawn at lunchtime. How does this stop heat coming in?

Kitchen heat transfers

How is heat transferred when cooking a meal?

Cooking a meal of soup and toast involves many heat transfers.

- Heat travels through the solid saucepan base by **conduction**. Saucepans made from metal are good heat conductors. The handle is often covered in plastic, a heat insulator, to stop you burning your hands.
- Heat spreads through the soup by **convection**. Soup is heated from the bottom of the pan. As hot soup rises, the cooler soup falls and is warmed.
- Heat **radiates** from the elements of a toaster to heat the bread. The inside of the toaster is shiny to reflect the heat onto the bread.

A

 1 a) Why are most saucepans made of metal?

b) Why are their handles often made of plastic?

2 Why won't a pan of soup heat properly under the grill?

Stopping heat losses

Kettles are designed to heat water effectively. The heating element is at the bottom so that convection currents heat the water quickly. A lid stops heat losses by convection. Jug kettles are made of plastic, a heat insulator, to reduce heat losses by conduction.

 3 Write down two ways that heat losses are stopped in kettles.

In some countries, meals are cooked in earth pits. Meat is heated up then wrapped and packed tightly into a pit lined with hay (a poor heat conductor). After several hours, the food is unwrapped, still hot and cooked through.

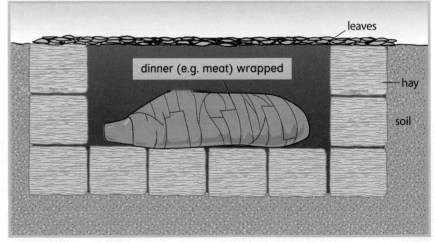

dinner (e.g. meat) wrapped

leaves

hay

soil

C

lid stops hot steam escaping

plastic is a poor heat conductor

water is heated from the bottom of the kettle

metal element conducts heat well to the water

B

 4 How are heat losses stopped in a cooking pit?

P What is the quickest way to cook a potato?

- Should you cut it up?
- Should you put a lid on the pan?
- Should you put a metal skewer through it?
- How would you find out which way is quickest?

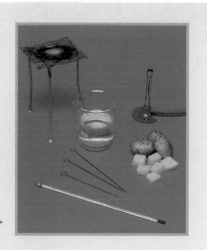

D

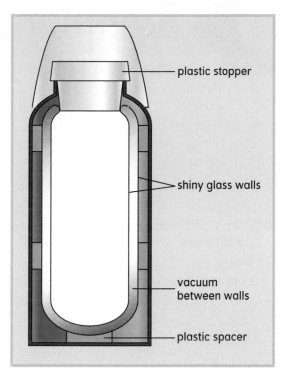

E *A vacuum flask.*

plastic stopper

shiny glass walls

vacuum between walls

plastic spacer

Vacuum flasks keep drinks hot (or cold) for many hours. They are designed to prevent conduction, convection and radiation.

? 5 Look at picture E. Copy and complete these sentences:

a) Plastic and glass are poor ____, and are known as ____.

b) The vacuum (space between the glass walls) stops ____.

c) The silver surface reflects heat back into the flask and stops ____.

! A cup of tea can still be hot enough to scald a baby 20 minutes after it is poured.

? 6 Jade is serving stew at a party from two containers. One is an open topped metal dish, the other is a plastic bowl with a lid. Explain which container keeps the stew hottest.

7 Why do kebabs cook quicker using a metal skewer instead of a wooden skewer?

8 Look at diagram F. How does the design of the barbecue help conduction, convection and radiation cook the food quickly?

F

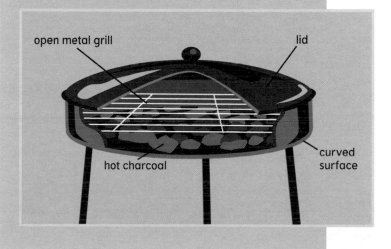

open metal grill

lid

hot charcoal

curved surface

Summary

Conduction, convection and ____ help us to cook food. Metal equipment helps ____ take place. ____ are used to prevent conduction. We use lids to stop hot air escaping and prevent ____. Solid food cooks by conduction, but ____ cook using convection.

conduction   convection   insulators
liquids   radiation

Stopping heat losses at home

Where is heat lost in the home and how can you stop it?

A lot of the energy used to heat our homes escapes. Cutting heat losses cuts down heating bills. Houses lose heat in several different ways:

- Most heat is lost by **conduction** through solid surfaces like floors, walls and windows.
- Warm air rises by **convection** through the house and is lost through the roof.
- Warm air is also lost in **draughts** through gaps in doors, windows and between floorboards.
- Surfaces like walls and windows **radiate** heat from the outside walls of a house.

Look at the two houses in diagram B.

Insulating the loft saves money. **A**

? 1 Where is most heat lost by conduction in the home?

B *One house is insulated and one is not.*

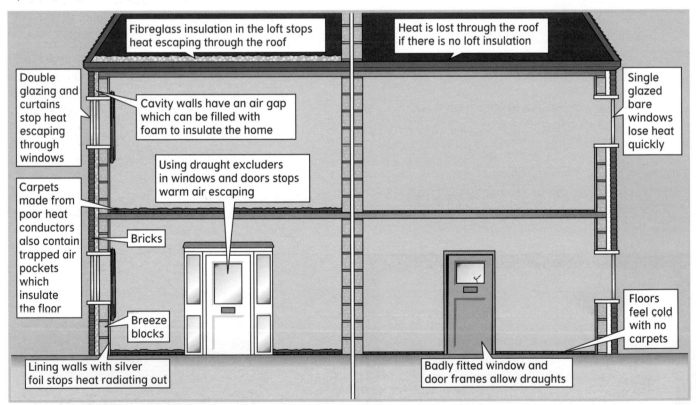

Fibreglass insulation in the loft stops heat escaping through the roof

Heat is lost through the roof if there is no loft insulation

Double glazing and curtains stop heat escaping through windows

Cavity walls have an air gap which can be filled with foam to insulate the home

Single glazed bare windows lose heat quickly

Using draught excluders in windows and doors stops warm air escaping

Carpets made from poor heat conductors also contain trapped air pockets which insulate the floor

Bricks

Breeze blocks

Floors feel cold with no carpets

Lining walls with silver foil stops heat radiating out

Badly fitted window and door frames allow draughts

? 2 Copy and complete table C to match each type of heat loss with one of the methods used to prevent it.

C

Type of heat loss	Method used to prevent it
Conduction	
Convection	
Radiation	

3 Look at diagrams D and E.

a) How does foam in the cavity wall help to prevent heat losses?

b) How does double glazing help to prevent heat losses?

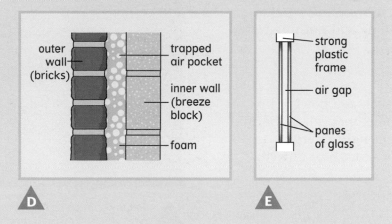

outer wall (bricks)
trapped air pocket
inner wall (breeze block)
foam

D

strong plastic frame
air gap
panes of glass

E

Cutting the cost of heating your home

It can be expensive to heat your home. Fitting draught excluders is cheap to do and can cut down heating bills a lot. However, double glazing is very expensive and does not cut fuel bills by much. To choose the best method, you need to find out

- how much it will cost to fit
- how much money you will save on your heating bills each year.

The best methods are cheap to fit, and save a lot of money each year. To be worthwhile, the amount that you save on heating bills over a few years should be more than you spend insulating your home.

P Insulating hot water tanks stops heat escaping into the house. How would you find out the best material to use? How thick should it be to keep the water warm? **F**

4 Explain whether each person below should fit double glazing or draught excluders.

a) Luca's house has small, badly fitting windows. He will move in 2 years.

b) Asif's house has large, well fitting windows. He plans to live there for 15 years.

! Swedish homes are so well insulated, they use only one-third of the energy used to heat British homes.

5 How can loft insulation help keep a house warmer?

6 Use your ideas about convection to explain

a) why central heating boilers are downstairs

b) why hot water tanks are upstairs.

Summary

Heat is lost from houses through walls, ____, windows and doors. Draughts occur when ____ air escapes through gaps. ____ ____ and closed curtains insulate windows. Carpets made from poor heat ____ insulate the floor using trapped ___ pockets. Using ____ in windows or ____ ____ on the walls stops losses by radiation. ____ walls insulate the house.

air   cavity   conductors   curtains   double glazing
floors   silver foil   warm

Types of energy
What different forms of energy are there?

Energy is needed for everything; without it, the Earth would be frozen, dark and silent. Stars would not shine and nothing would move. Life would not exist.

Energy in action

There are many types of energy. Objects such as a bowl of hot soup have **heat energy**. Heat energy is sometimes called **thermal energy**. Objects that glow or shine give out **light energy**. Objects that give out light energy often give out heat as well.

A *There would be no life without heat and light energy from the Sun.*

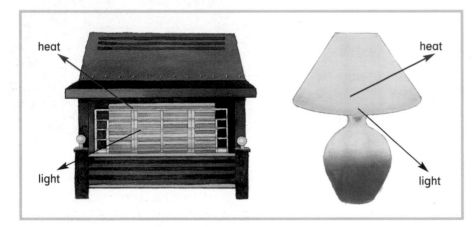

? **1** A fire gives out heat and light energy. Name two other things that give out heat and light energy.

B *Both items give out heat and light.*

Anything making a noise gives out **sound energy**. Sometimes you can feel the vibrations caused by very loud sounds.

? **2** A barking dog gives out sound energy. Name two other things which give out sound energy.

All moving objects have energy called **kinetic energy**. A fast car and a moving snail have kinetic energy. Faster objects have more kinetic energy.

? **3** A running child has kinetic energy. Name two other things that have kinetic energy.

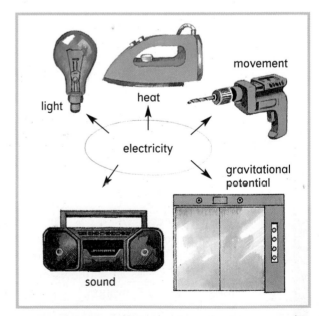

C

Electrical energy is used by anything that plugs into the mains or uses electrical cells (batteries). Electrical energy can be changed into many other types of energy. Our lives would be very different with no electricity.

Energy is measured in **joules (J)**

? **4** Name two things which use electrical energy

Stored energy

Some types of energy are stored. Anything that can fall down has **gravitational potential energy**. Another type of stored energy is **elastic potential energy**, which is found in objects that are stretched or squashed, such as a stretched elastic band.

Chemical energy is a type of stored energy found in food and fuels. For example a lump of coal and a plate of chips both contain chemical energy.

Summary

____ is needed for anything to happen. There are five types of energy in action: heat, ____, ____, ____, and ____. Energy can also be stored as ____, ____ ____ and elastic potential energy.

chemical   electrical
gravitational potential   kinetic
light   sound   energy

5 A person climbing stairs gains gravitational potential energy. Name two other things that gain gravitational potential energy.

6 A blown up balloon has elastic potential energy. Name two other things that have elastic potential energy.

P Machines can give out several different forms of energy. What different forms of energy does each piece of equipment give out?

D

7 A bar of chocolate contains chemical energy. Name two other things that contain chemical energy.

8 What forms of energy do these things give out?

a) a radio   **b)** the Sun   **c)** a television.

9 Look at photograph F. For each form of energy, write down three places in the kitchen where it is being used or stored:

a) gravitational potential energy   **b)** chemical energy
c) heat energy   **d)** electrical energy.

F

Using potential energy

How can we use potential energy?

Energy is stored in some objects because of their position or shape. This stored energy is called **potential energy**. Because potential energy can be changed easily into different forms of energy, it is used in many different places.

? 1 What do we mean by potential energy?

As roller coaster carriages are lifted to the start of the ride, they gain **gravitational potential energy**. This rapidly changes into kinetic (movement) energy when the ride starts and the carriages drop down the track. Anything that can fall down has gravitational potential energy.

Objects have more gravitational potential energy if they are

- higher up
- heavier (weigh more).

? 2 How can an object gain gravitational potential energy?

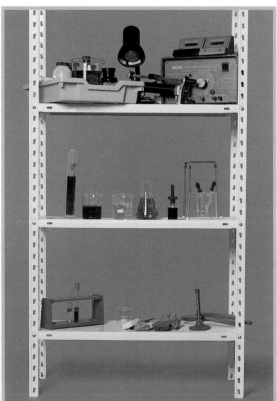

 Heavy items on the top shelf store more gravitational potential energy than items on lower shelves.

A

E Gravitational potential energy can be calculated if you know

- the object's weight (in newtons)
- the height it has moved up or down (in metres)

change in gravitational potential energy = **weight** × **change in height**
(in **joules, J**)   (in **newtons, N**)   (in **metres, m**)

Example

Nick weighs 600 newtons. How much gravitational potential energy does he lose when he drops 10 metres on a roller coaster ride?

- Nick's weight is 600 N.
- His change in height is 10 m.
- The gravitational potential energy he loses is

  600 N x 10 m
  = (600 x 10) J
  = 6 000 J

P How would you use your ideas about gravitational potential energy to make a roller coaster ride faster and more exciting?

c

Summary

Potential energy is _____ energy that can be used later. If an object is raised up, it stores gravitational _____ energy. More energy is stored if _____ objects are lifted, or they are lifted _____. You can calculate gravitational potential energy using this equation:

gravitational potential energy (J) =
_____ (N) x change in vertical _____ (m)

| heavy | height | higher | potential | stored | weight |

?

3 Sue weighs 500 newtons. She dives 4 metres into a swimming pool.

  a) What is Sue's weight?
  b) What is her change in height?
  c) How much gravitational potential energy does she lose?

4 Explain which object has more gravitational potential energy:

  a) a football just as it is kicked or just as it is headed?
  b) a car or a lorry at the top of a multi-storey car park?

5 **a)** Explain where a child on a slide has the most gravitational potential energy.
  b) What form of energy does gravitational potential energy change into as the child slides down?
  c) Which other playground equipment changes gravitational potential energy into kinetic energy?

6 Work out the gravitational potential energy that these objects have:

  a) A bird weighing 1 N, perched at the top of a 20 m high tower.
  b) A football weighing 7 N, kicked 3 m up into the air.

179

Energy transfers

How does energy change from one form to another?

We need energy to do everything. The same is true for machines. Diagram A shows how electrical energy from electrical cells can be changed into many other forms of energy. **Electrical energy** is useful because it can be changed into so many other forms of energy.

A

As the float climbs the hill, it gains gravitational potential energy

The horn gives out sound energy

The moving float has kinetic energy

milk float

Flashing indicators give out light energy

Electrical cells are charged up using electrical energy from the mains

Forms of unwanted energy from the engine include heat and sound

Q286 JKR

Unigate

1 In the milk float, where is electrical energy changed into

   a) movement
   b) heat
   c) light
   d) sound?

2 Which piece of equipment can store energy from the electrical supply?

! 100 million joules of electrical energy changes into heat energy in one lightning strike.

Some forms of energy can be stored for later use. The milk float's electrical cells can be charged up from mains electricity. The electrical energy is changed into chemical energy, which is stored in the cells.

This energy transfer can be written like this:

electrical energy   →   chemical energy

When the milk float is used, chemical energy is changed back into electricity.

chemical energy   →   electrical energy

Diagrams like these are called **energy flow diagrams**.

Changing electrical energy

Electricity is a big part of all our lives. Many pieces of equipment use electricity because it is so convenient and clean.

At the flick of a switch, a kettle changes electrical energy into heat energy. A TV changes electrical energy into light and sound energy. As you go up in a lift, electrical energy is changed into kinetic (movement) energy and gravitational potential energy.

P Electrical energy can be transferred into at least four different types of energy. How would you find out which piece of equipment can carry out the most transfers?

C

B

?

3 Picture B shows an electrical goods shop. There are many different electrical appliances. Name one appliance that changes electrical energy into heat energy.

4 Write down the energy transfers that take place when a washing machine is being used.

?

5 Choose two pieces of kitchen equipment and write down their energy transfers.

6 a) Copy and match these items with the energy transfer they perform:

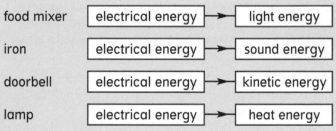

food mixer — electrical energy → light energy

iron — electrical energy → sound energy

doorbell — electrical energy → kinetic energy

lamp — electrical energy → heat energy

b) Are any other forms of energy given out by these items of equipment?

8 Asha switched off her clock radio and crawled out of bed. She turned on the light, showered quickly, and then dried her wet hair with her hair dryer. She had some coffee and toast while listening to a CD, she then ran for the bus.

a) Write down all the electrical equipment Asha used.

b) For each item, write down the type of energy the electrical energy was turned into.

Summary

Energy can be transferred from one _____ into another. _____ energy can be changed into heat, light, _____, kinetic and gravitational _____ energy.

| electrical | form |
| potential | sound |

Power

What is power?

Sound Barrier Broken

Black Rock Desert, Nevada. 15 October 1997. Andy Green breaks the sound barrier, travelling at 1228 kilometres per hour, in his car *ThrustSSC*.

The car is the brainchild of Richard Noble and built to travel faster than most aeroplanes. It uses Rolls Royce jet engines to provide a huge amount of energy every second and allow the car to hurtle along at supersonic speeds.

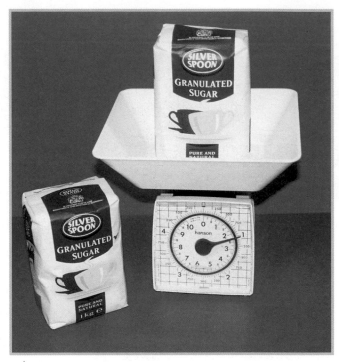

B *Lifting the sugar gave it 1 extra joule of gravitational potential energy.*

A

? **1** Why can *ThrustSSC* travel faster than most aeroplanes?

Energy is needed before anything can happen. Chemical energy stored in *ThrustSSC's* fuel made it move. The *ThrustSSC* used up the energy from its fuel very quickly because it had very powerful engines. **Energy** is measured in **joules (J)**. One joule is a tiny amount of energy. Lifting a bag of sugar 10 cm uses up 1 joule of energy. *ThrustSSC* used 60 million joules each second!

The **power** of a machine measures how quickly its energy is supplied, or used up. Power is measured in **watts (W)** or **kilowatts (kW)**. There are 1000 watts in a kilowatt. A watt is a tiny amount of power — a very dim light bulb has a power of 25 watts.

? **2** What is the unit of power?

E We can measure the power of equipment by finding out how quickly it uses up energy. Power is calculated using this equation:

power	=	**energy**	÷	**time**
(in **watts, W**)		(in **joules, J**)		(in **seconds, s**)
power	=	energy *divided by*		time

More powerful items use up more energy every second.

A nightlight bulb has a power of 25 watts. It uses 25 joules of energy per second and gives out very dim light.

A bright reading light uses a 100 watt bulb, so it uses up 100 joules every second.

The dim light only uses a quarter of the energy of the bright light each second.

3 Write down the equation we use to calculate power.

4 What is *ThrustSSC*'s power?

P

You can find out how powerful you are by measuring how many step-ups you can do in 1 minute.

D

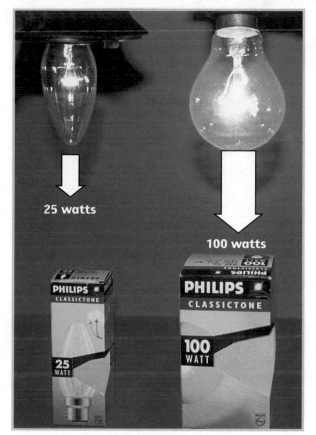

25 watts

100 watts

C

The Space Shuttle burns 98 tonnes of fuel in 1 minute and has a power of 17 600 kW.

5 How much energy does a 150 watt fridge use each second?

6 What is the power, in watts, of a 2 kW bar heater?

7 How much energy does a 1.2 kilowatt hairdryer use each second?

8 Which is most powerful: a 2 kW kettle or a 1500 W iron?

9 Copy and complete table E:

E

Item	Power (watts)	Power (kilowatts)	Energy used in 1 second (joules)
Dishwasher		2 kW	
Electric blanket	100 W		
Television			300 J
Vacuum cleaner			1.5 kJ

10 How much energy does a mobile phone charger transfer in 1 hour? Its power is 6 W.

Summary

Energy is measured in _____ or kilojoules.
Power measures how fast _____ is transferred.
Power is measured in _____ or kilowatts. A kilowatt is a _____ watts.

Power (W) = $\dfrac{\text{energy (J)}}{\text{time (s)}}$.

| energy | joules |
| thousand | watts |

Electrical energy and power

How much energy do electrical devices transfer?

Clubs have powerful loudspeakers to pump out loud music. These loudspeakers use a lot of electrical energy each second, to give out so much sound energy. The equipment is on for a long time so a great deal of electrical energy is used.

 1 What energy transfer takes place in the loudspeakers?

E Hi-fi equipment at home is not as powerful as a club's equipment. The amount of electrical energy used at home and at the nightclub can be worked out if you know:

● the power of the equipment (in watts) and
● the time it is on for (in seconds).

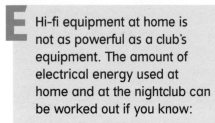
energy transferred = power × time
(in **joules, J**)   (in **watts, W**)   (in **seconds, s**)

Remember: multiply the number of minutes by 60 to change them into seconds.

A

Example

A radio (power 100 W) is on for 1 minute (or 60 seconds). How much energy is used?

● Its power is 100 W.
● It is on for 60 seconds.
● Energy = power × time
  = 100 W × 60 s
  = 6000 J.

 2 A 650 watt microwave oven is on for 30 seconds.

a) What is its power?
b) How long is it on for?
c) How much energy does it use?

Electrical appliances can be powerful and are sometimes left on for a long time.

3 Look at picture B.

a) What is the power of this microwave oven?
b) How much energy does it use if it is switched on for 2 minutes?

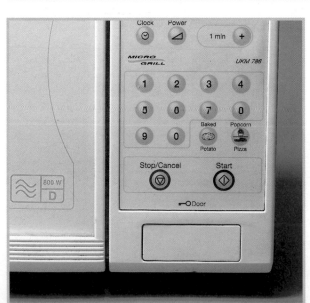

B

P How would you find out if a kettle transfers the same amount of energy in each second that it is switched on?

E The energy used by more powerful electrical equipment can be measured in **kilowatt-hours**.

You need to know:

- the power of the equipment (in kilowatts)
- the time it is on for (in hours).

energy transferred	=	**power**	×	**time**
(in **kilowatt-hours**, **kWh**)		(in **kilowatts**, **kW**)		(in hours, **h**)

Remember: Divide the power in watts by 1000 to change it into kilowatts.

Example

An iron (power 2 kW) is on for 1 hour. How much energy is used?

- The power of the iron is 2 kW.
- It is on for 1 hour.
- Energy = power × time

$$= 2 \text{ kW} \times 1 \text{ h}$$
$$= 2 \text{ kWh}.$$

? **4** How much energy is used if the iron in the example above is on for 2 hours?

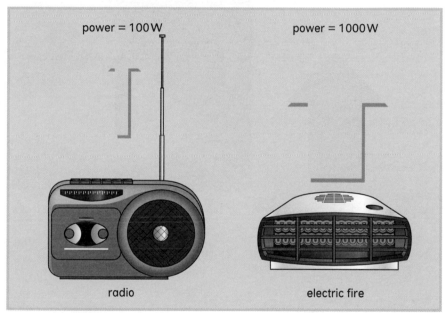

power = 100 W   power = 1000 W

radio   electric fire

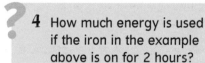

The radio uses 1 kilowatt-hour of energy in 10 hours. The more powerful heater only takes 1 hour to use 1 kilowatt-hour.

? **5** The power of an oven is 5 kW.

    a) How much energy (in kWh) does it use in 1 hour?

    b) What is its power in watts?

    c) How much energy (in joules) does it use in 10 seconds?

6 Ross watched TV (0.3 kW) for 3 hours one night. His mum used the tumble dryer (4 kW) for half an hour. Who used most energy?

Summary

Energy is measured in joules or in _____ _____. We can calculate the energy transferred using:

energy (J) = power (W) × time (_____). Or
energy transferred in kilowatt-hours (kWh) = power (kW) × time (_____).

kilowatt-hours   s   h

Buying electricity

How can you calculate the cost of electricity bills?

The Jacob family's electricity bill in December was much larger than September's bill. Mr Jacob thought that the electricity company had put up its prices. Then he saw that more Units of electricity had been used. The lights were on in the dark evenings and so were the heaters.

 1 Why was Mr Jacob's electricity bill bigger in winter?

The amount of electricity used in homes is measured in **kilowatt-hours**, often just called a **Unit**. More Units are used if

- more equipment is on
- more powerful equipment is on
- the equipment is on for a longer time.

Remember:
energy (in Units or kWh) = power (in kW) $\times$ time (in h).

 The total cost of the electricity used can be calculated if you know:

- the number of Units (or kilowatt-hours) used and
- the cost per Unit (in pence).

total cost = number of Units used $\times$ cost per Unit
(in **pence**)   (in **pence**)

 How would you find out which item of electrical equipment increases your electricity bill the most?

B

A

 2 What is a Unit of electricity?

Example

The Jacob family used 1280 Units. Each Unit cost 5.9 pence.

The cost of electricity used was:

1280 Units x 5.9 p

= 7552 p (or £75.52).

The electricity meter shows how many Units have been used. Every three months an electricity bill is prepared, showing the last meter reading and the present meter reading. The number of Units to be paid for is the difference between the two readings.

Example

The Jacob's bill was prepared using these readings:

- last meter reading: 76885 Units
- present meter reading: 78165 Units
- number of units used = 78165 − 55542
  = 1280 Units.

3 Look at the electricity bill in picture C.

a) How many Units were used?

b) How much did each Unit cost?

c) What was the total amount to pay?

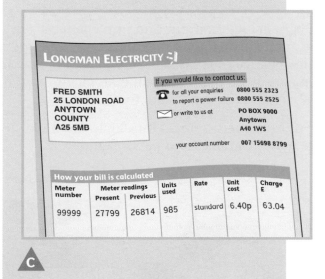

4 Look at the meter in pictures D and E.

a) What was the reading in May?

b) What was the reading in August?

c) How many Units have been used between May and August?

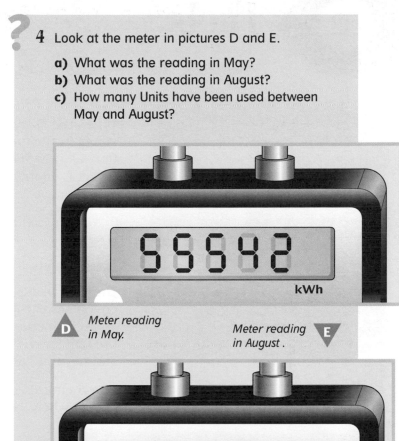

D Meter reading in May.

E Meter reading in August.

Summary

The amount of electricity used depends on the _____ of the equipment used and how long it is on for. It is measured in _____ -hours, or Units.

Energy used (Units or kWh) =
power (kW) × _____ (h)

Total cost (pence) =
number of _____ used × cost per Unit (pence)

_____ of Units used =
present reading − previous reading

kilowatt	number	power
time	Units	

5 Look at the readings in Table F.
Calculate the number of Units used.

Month	Last reading	Present reading	Units used
February	52045	53499	
May	53499	54648	

F

6 How much did the electricity cost for these bills?

a) 1056 Units costing 6.4 p each.

b) 1103 Units costing 6.4 p each.

Fuels

What are the different types of fuels?

A

When the Space Shuttle takes off, fuel burns. Chemical energy stored in the fuel is changed into kinetic energy, forcing the rocket up and away from the Earth. The energy needed to launch the rocket is enormous and all the fuel is used up within 2 minutes.

1 What sort of energy is stored in fuels?

Fossil fuels

Growing plants turn light energy from the Sun into chemical energy. After they die, many plants become buried. Over millions of years, heat and pressure change the buried dead plants into **coal**. In the same way, the remains of sea plants and animals change into **oil** and **natural gas**. Coal, oil and natural gas are **fossil fuels**.

The chemical energy in fossil fuels changes into light and heat energy when they are burnt in homes and power stations.

B *From this...*   *....to this in 200 million years.*

plants

sea creatures

coal

oil

Oil is used for many things. **C**

2 Unscramble the letters to name these fossil fuels:

oli   loca
an atlas rug.

3 Choose the right answer. Fossil fuels are used to

A) decorate the home
B) heat and light homes
C) insulate homes.

P How would you compare the different amounts of energy stored in different fuels?

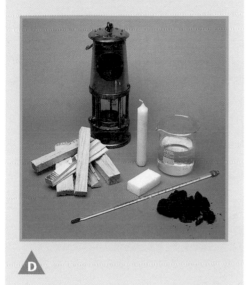

D

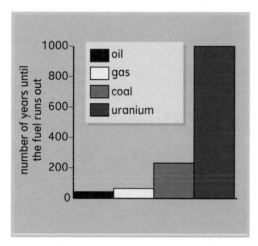

F *Many of the fuels we use will run out.*

number of years until the fuel runs out

- oil
- gas
- coal
- uranium

1000
800
600
400
200
0

5 Why is nuclear fuel non-renewable?

6 Why is wood a renewable energy resource?

7 Why will some fuels run out one day?

8 Write a paragraph about how your life would change without coal, oil and natural gas.

After fossil fuels have been burnt only ashes remain, and these cannot be re-used. Fossil fuels are **non-renewable** energy resources, which means that they cannot be re-used or replaced. It takes millions of years for fossil fuels to form, and we are using up fossil fuels much faster than they are produced. They will run out soon unless we use other ways of providing energy for our everyday needs.

4 Copy and complete this sentence. A non-renewable energy resource cannot be ...

E *Once it has been burnt, coal cannot be re-used.*

Other fuels

Nuclear power stations use **uranium** and **plutonium** as fuels. Nuclear reactions release energy from these fuels. Supplies of these metals are limited. Once they have been used up, they cannot be replaced. Nuclear fuels are non-renewable.

Wood is another fuel, although it gives out less energy than fossil fuels when it burns. Fast growing trees planted now can be used for fuel in 10 years time. Wood is a **renewable energy** resource because we can replace the trees that are cut down.

Summary

Fossil fuels give out energy when they are ____. They take millions of years to produce and once burnt, they cannot be ____ so they are non-____ fuels. Coal, ____ and natural ____ are ____ fuels. Other non-renewable fuels include uranium and plutonium, which are ____ fuels. Non-renewable energy resources will last ____ if we use them more slowly. ____ can be replaced so it is a renewable fuel.

| burnt | fossil | gas | longer | nuclear |
| oil | replaced | renewable | wood | |

Fuels and the environment

How do fuels affect the environment?

Severe storms and floods can cause deaths and damage to buildings and crops. Some scientists think more storms and floods occur now because we burn too many fossil fuels.

Fossil fuels are a good energy source because:

- their energy is very concentrated
- it is easy to transport fossil fuels to power stations.

However, burning fossil fuels adds carbon dioxide to the air. There is more in the air now than there was one hundred years ago. High carbon dioxide levels trap some of the Sun's heat in the atmosphere. This is called the **greenhouse effect**. The greenhouse effect is making the temperature of the atmosphere increase slowly. This is called **global warming**.

heat and light from sun

some heat is absorbed by gases like carbon dioxide

atmosphere

some heat is reflected back to earth by gases like carbon dioxide

A *The greenhouse effect.*

 1 Choose the correct answer.
Global warming is caused by …

A) the Sun getting closer
B) more carbon dioxide in the air
C) more greenhouses.

 Severe floods in Bangladesh have killed tens of thousands of people. As sea levels rise due to global warming, many more could die in the future.

 2 Why do we use fossil fuels?

Burning coal, oil and natural gas releases **pollution** into the air. Sulphur is found in fossil fuels. When it burns it forms sulphur dioxide. This dissolves in the clouds, forming acid rain. Acid rain can kill plants and animals living in lakes and ponds.

The sulphur can be removed from fuels before they are burnt, or sulphur dioxide can be removed from waste gases. However, this makes the fossil fuels more expensive to use.

 The effects of acid rain

3 Choose the correct answer.
Acid rain is caused by

A) sulphur dioxide
B) carbon dioxide
C) oxygen.

In 1989, school children in Mexico City had a month off school because pollution caused such severe smog.

Type of pollution	Effect on environment	Ways to reduce effect
Carbon dioxide	It increases the greenhouse effect, causing global warming.	Burn less fuel.
Sulphur dioxide	It dissolves in rain making it acidic. **Acid rain** kills plants and fish.	Remove the sulphur from the fuel before it is burned. Fit filters in power station chimneys to absorb the sulphur dioxide produced.

Different fuels release different amounts of carbon dioxide for the same amount of energy released. Natural gas releases the least, then oil, and coal releases the most.

4 Put these fuels in order starting with the fuel releasing the most carbon dioxide: oil, natural gas, coal.

Nuclear power

Nuclear power does not cause global warming or acid rain, but it creates **radioactive waste**. Fairly small amounts are produced which are sealed in glass and stored underground. Nuclear waste needs careful storage because it can stay radioactive for thousands of years. A **nuclear accident** can release a lot of radioactivity into the surrounding areas. In Britain, strict controls on reactor design make nuclear accidents very unlikely.

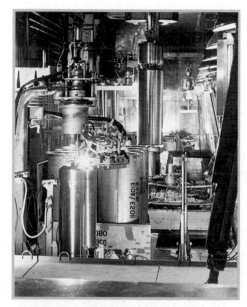

D *Nuclear waste can be stored in glass then sealed in metal drums.*

5 How is radioactive waste made safe?

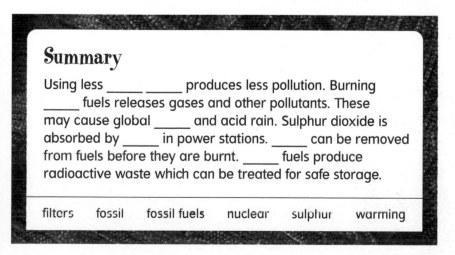

Summary

Using less _____ _____ produces less pollution. Burning _____ fuels releases gases and other pollutants. These may cause global _____ and acid rain. Sulphur dioxide is absorbed by _____ in power stations. _____ can be removed from fuels before they are burnt. _____ fuels produce radioactive waste which can be treated for safe storage.

filters   fossil   fossil fuels   nuclear   sulphur   warming

6 Work out these muddled up words:

a) it is used to absorb sulphur dioxide (trifle)
b) it creates radioactive waste (pure crown ale).

7 One way to create less pollution is to produce less electricity. Design a poster to persuade your friends to use less electricity.

Power stations

How is electricity created from energy stored in fuels?

Electricity is generated in power stations. Many power stations change kinetic energy into electrical energy using huge magnets and coils of wire.

 1 What things are needed to generate electricity?

Where kinetic energy comes from

Energy stored in fuels is used to heat water until it turns into steam. The steam gains the kinetic energy that is needed to generate electrical energy. Power stations use:

- fossil fuels like coal, natural gas and oil
- the nuclear fuels uranium and plutonium.

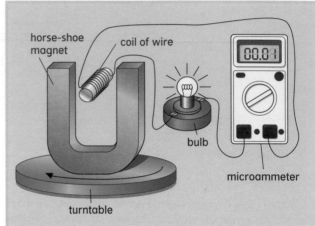

A

 2 What are fuels used for in a power station?

Why steam is needed in power stations

The steam has enough kinetic energy to turn the blades of large **turbines**.

The turbine spins a magnet, which is inside a coil of wire. The magnet and coil of wire is called a **generator**.

The kinetic energy of the turbine changes into electrical energy inside the generator.

This makes an electric current flow in the coil of wire.

A turbine.

 3 Copy and complete this sentence. A turbine is used to ____ a magnet inside the ____.

P How would you make your own electricity generator from magnets and wire?

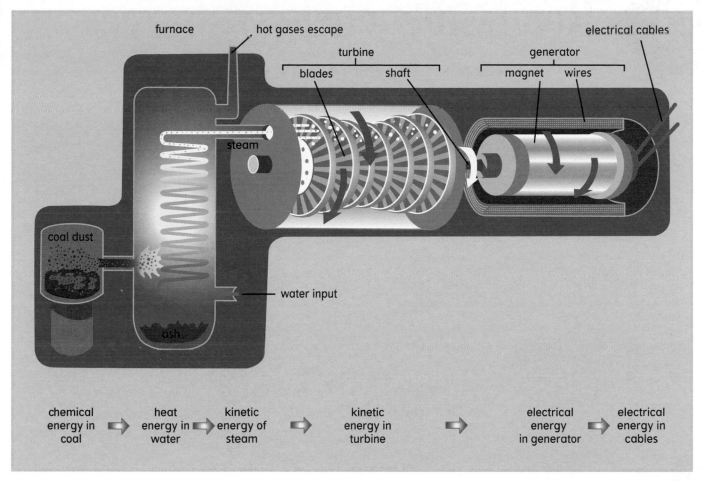

furnace · hot gases escape · electrical cables

turbine · generator

blades · shaft · magnet · wires

steam

coal dust

water input

ash

| chemical energy in coal | | heat energy in water | | kinetic energy of steam | | kinetic energy in turbine | | electrical energy in generator | | electrical energy in cables |

Diagram C shows what happens inside a coal-fired power station.

 C

! An average person in India uses 50 times less energy per day than an average person in the USA.

? 4 Copy this energy flow diagram. Fill in the missing energy changes that happen inside the power station:

| _____ energy |
↓
| _____ energy |
↓
| _____ energy |
↓
| _____ energy |

? 5 Explain why water is turned into steam in a power station.

6 Explain how steam helps the turbine to spin.

7 Why is electrical energy not used to power the station?

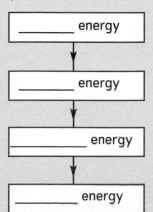

Summary

Power stations use _____ to heat water and change it into steam. Steam turns _____, which spin large magnets. Magnets spinning inside a coil of wire produce an _____ current.

electric   fuels   turbines

Renewable energy sources

What is a renewable source of energy?

For centuries, grains of corn and wheat have been ground into flour for cooking. Kinetic energy from flowing rivers helped to turned huge stones in watermills. Windmills used kinetic energy from the wind.

The energy from flowing water and the wind are **renewable sources of energy**, which will never be used up.

? **1** Copy and complete this sentence.
A renewable source of energy will …

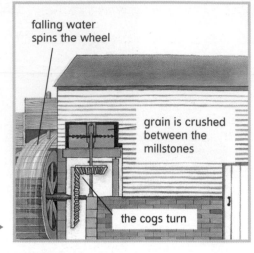

falling water spins the wheel

grain is crushed between the millstones

the cogs turn

A

Gravitational potential energy can be stored in reservoirs for later use. **B**

Hydroelectric power

When rivers flow into a lake behind a dam, they fill reservoirs with large amounts of water.

- The water has gravitational potential energy.
- This energy changes into kinetic energy when the water falls through pipes.
- The pipes contain turbines which spin as the water rushes through.
- The turbines spin generators.
- The generators change the kinetic energy into electrical energy.
- This is called **hydroelectric power**.

? **2** Copy this energy flow diagram. Complete the missing energy transfers inside the hydroelectric scheme:

| ____ energy | → | ____ energy | → | ____ energy |

P How would you investigate what is the best design for a waterwheel? Test your idea by using it to lift a small mass.

! Dams can be dangerous. Tens of thousands of people were killed in India when a dam burst in 1979.

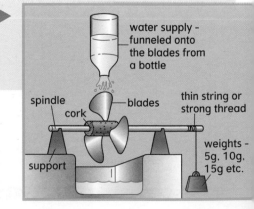

C

water supply – funneled onto the blades from a bottle

spindle
cork
blades

thin string or strong thread

support

weights – 5g. 10g. 15g etc.

Tidal power

Energy is available from moving water in the sea. Twice a day, huge quantities of water flow in and out of the estuaries as **tides** go in and out.

- Dams, called **barrages**, trap water at high tide.
- After the tide goes out, gates in the barrage are opened.
- Water rushes out through pipes, spinning turbines in the pipes.
- These turbines turn the generators, creating electricity.
- When the tide comes in, water rushes in and the turbines spin the other way.

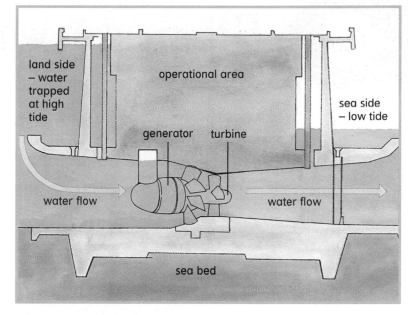

 3 What are the turbines in the pipes used for?

 The gravitational pull of the Moon and Sun provides the energy needed to produce electricity!

Problems

Hydroelectric power and tidal power both use **renewable** energy resources. Although no pollution is produced, there are problems.

In hydroelectric schemes, whole valleys are flooded using large dams, destroying farming and forestry land.

In tidal schemes, huge barrages across river mouths are needed, permanently flooding large areas of river estuaries. This detroys habitats of wading birds and other wildlife. Very few places have tides that are strong enough for a tidal scheme.

 Tidal barrage in France.

Summary

Renewable energy resources will not ____ ____. Running water and tides can spin ____ which turn a generator to produce electricity. No ____ is caused by hydroelectricity or tidal schemes but they both ____ large areas.

pollution   flood   run out
turbines

4 Write down two ways that falling water is used to produce electricity.

5 In a tidal scheme, describe where

  a) gravitational potential energy is changed into kinetic energy

  b) kinetic energy is changed into electrical energy.

6 Give one advantage and one disadvantage each for tidal schemes and hydroelectric power.

7 Imagine that the Government want to build a tidal power scheme near you. Write a letter to your MP to explain whether or not you think it is a good idea.

Other renewable sources

What other renewable sources of energy can we use?

The world's population is increasing. Technology is widespread and travel is easy but fossil and nuclear fuel supplies are running out. We have only used electricity for 100 years, but already billions of people depend on it. Using renewable energy resources like the wind, waves and the Sun will help our fossil and nuclear fuels last longer.

 1 How can we make our fuels last longer?

Wind power

When a strong wind blows, the fast moving air turns large blades on **wind turbines**. This spins the generator, producing electricity. Groups of wind turbines work best in windy places like hilltops and coastlines so that the wind can turn the blades easily. There are problems though:

- Many people think wind turbines are ugly and they are noisy. They cause visual and noise pollution.
- The amount of electricity produced changes with the amount of wind.
- If it is calm, no electricity is produced.
- Wind turbines must stop in storms to prevent them being damaged.

 2 What affects how much electrical energy a wind generator produces?

A A group of wind turbines like this is called a wind farm.

P How would you find out what is the best design for a wind turbine? The electricity is produced using a small generator (called a dynamo). **C**

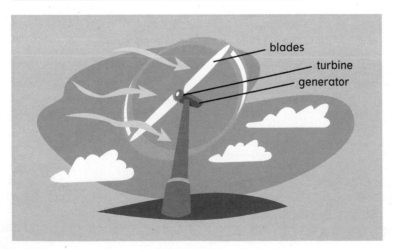

blades
turbine
generator

B

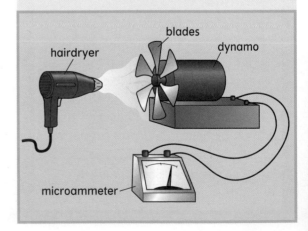

hairdryer
blades
dynamo

microammeter

Geothermal power

In some volcanic areas, steam is produced naturally underground. It is piped when it reaches the surface, and used to turn turbines directly. This renewable energy resource is called geothermal power. The underground rocks are hot because radioactive elements in them, including uranium, produce heat as radioactive decay occurs.

Wave power

It is possible to get energy from **waves**. Small rafts move up and down with the water. This movement can be used to turn small turbines, which generate electricity. However, the amounts of electricity which can be produced at the moment are too small to be useful.

Solar power

When sunlight shines on **solar cells**, they convert energy into electricity. Solar cells are useful:

- in sunny, remote places
- on satellites many miles away from Earth
- if very small amounts of electricity are needed, e.g. in watches or calculators.

3 Where is solar power useful?

There are some problems with solar cells:

- they are expensive
- the amount of electricity produced depends on the amount of sunlight
- if it is not sunny, little or no electricity is produced.

4 Why will fossil and nuclear fuels last longer if we also use renewable energy resources?

5 Copy and complete this sentence. Inside a wind turbine, _____ energy changes into _____ energy.

6 John thinks that all electricity should be generated using renewable resources. Write a few sentences explaining the problems this could cause.

D *Thousands of machines like these are needed to generate the same amount of energy that is supplied from one fossil fuel power station.*

E

Summary

Renewable energy resources include waves, geothermal power, the _____, and the Sun. These renewable resources do not cause so much _____ but large wind farms are _____ and many people think they are ugly. Wind and wave power does not produce much _____. Solar cells are used in _____ places.

electricity   noisy   pollution
sunny   wind

Fuels for electricity

What is the best method for producing electricity?

The amount of energy that we use can vary. In summer, less energy is needed for heat and light. However, every evening, many people cook meals, watch TV and put the kettle on. TV schedules and the time of day provide important information for power station engineers so that they can control the amount of electricity available.

 1 Write down two things that affect how much electricity we use.

The best energy resource to use:

- does not pollute or affect the environment
- is plentiful, cheap and local
- provides the energy required when we want it.

The best power stations are:

- cheap to build
- safe to use
- easy and quick to stop and start.

In reality, no energy resource provides all of these. In the UK, we use a variety of energy resources.

 2 Write down the features of a good energy resource.

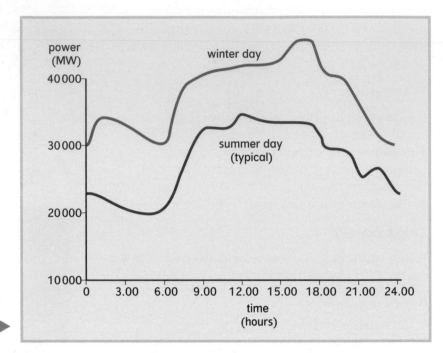

A

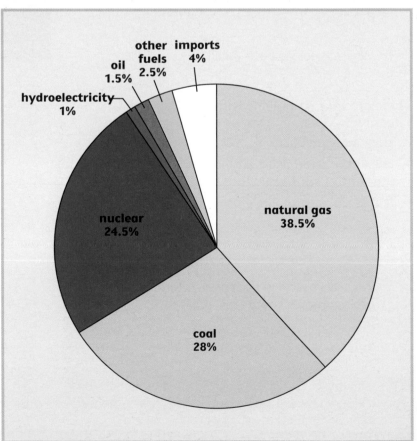

How the UK's electricity was provided in 1999. B

Each type of power station has advantages and disadvantages

Fossil fuel power stations:

- can use coal which is cheap and plentiful.
- can use natural gas efficiently.
- can use oil, but oil is expensive and oil reserves have other important uses so oil is not used much.
- can take hours to start and stop, so they cannot provide extra electricity quickly. Natural gas stations can start the quickest, then oil, and finally coal-fired stations.

Nuclear power stations:

- run all the time because it takes so long to stop and start them. It takes much longer to stop and start a nuclear power station than a fossil fuel power station.

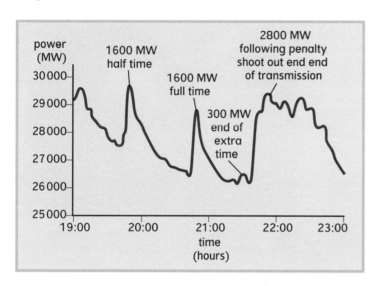

C *The dots in the top right of the screen warn engineers to expect a surge in demand during adverts.*

? **3** List these energy sources in order, starting with the one that is quickest to start generating electricity: nuclear, oil, coal, gas.

D *The demand for electricity during the 1990 England-Germany World Cup semi-final. Kettles and lights being turned on caused the peaks in demand.*

Power stations using renewable sources:

- Hydroelectricity provides electricity very quickly. The stations are used for surges in demand, for example during TV adverts when kettles are switched on!
- Hydroelectricity schemes can use spare energy produced by other power stations to pump water back into its reservoirs.
- Tidal schemes can provide energy in minutes if the tide is at the right part of its cycle. This varies with the time of day. The height of the tide varies too, depending on the month and the time of year.

Summary

Fossil and _____ fuels provide energy when it is needed but the power stations take a long _____ to start and stop. Renewable energy resources can't always provide energy when it is _____ but can start and stop in _____.

minutes   needed   nuclear   time

? **4** Write down one reason why fossil and nuclear fuels provide most of our energy.

5 Write sentences explaining why
   a) oil is not used all the time
   b) tidal schemes are not used for sudden surges in demand
   c) nuclear power is used all the time.

Energy losses

How is energy wasted?

Two friends are out for a cycle ride. Josh's bike is brand new and its pedals turn easily, the wheels go round quietly and the chain doesn't rub. Sam's bike is older and much harder to ride. The wheels get hot as they rub against the brake pads, the pedals squeak as they turn, and the chain catches as he rides.

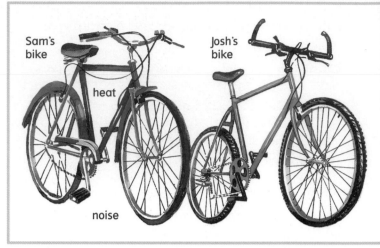

Sam's bike   Josh's bike

heat

noise

A

1 While Sam's bicycle is moving, it releases two other forms of energy. What are they?

Whenever energy is transferred, only part of it ends up in the form that is wanted. Some energy is wasted in the form of unwanted sound or heat. Eventually the wasted energy spreads out to the surroundings. It is very hard to re-use wasted energy once it has spread out.

2 What happens to the wasted energy?

! Water at the bottom of a waterfall is slightly warmer than at the top. However, this heat energy cannot be used as the temperature difference is very small.

B When the balloon pops, its elastic potential energy is lost to the surroundings as sound.

Wasted energy is usually in the form of **heat** or **sound**. For example, washing machines and vacuum cleaners make a noise while they are being used.

Other items that make a noise include cars and lorries, squeaky hinges and banging doors. You can feel warmth from computers and lamps when they are switched on. Car engines and tools like saws and drills become hot when they are used.

heat

chemical energy

sound

kinetic energy

C

3 List two forms of unwanted energy in a motorbike.

P How many forms of unwanted energy can you identify when these objects are used?

We can write the energy transfer from an electric drill like this:

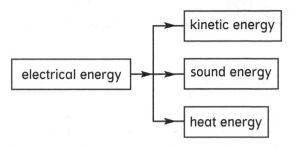

electrical energy → kinetic energy

electrical energy → sound energy

electrical energy → heat energy

We need kinetic energy from the drill. The wasted forms of energy are sound and heat.

The energy transfer from a lamp can be written this way:

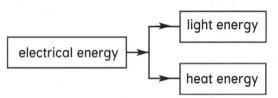

electrical energy → light energy

electrical energy → heat energy

?
4 Which form of energy is wanted from a lamp?

5 Which forms of energy are wasted from a drill?

Summary

When energy is transferred, some is changed into ____ forms. The most common forms of wasted ____ are heat and ____. Wasted energy is transferred to the surroundings, which ____ up.

energy   heat
sound   wasted

?
6 Copy and complete table F.
The first example has been done for you.

F

Item	Forms of energy given out	Wanted energy	Wasted energy
lamp	heat, light	light	heat
hairdryer			
bicycle			
lawnmower			
kettle			
bus			

7 Write down two ways to cut down on wasted energy transfers on a bicycle.

Efficiency

What is efficiency?

A **Fluorescent lights in a supermarket.**

The light bulbs we use at home get very hot when they are on. Some of the electrical energy changes into light energy, but a lot changes into heat. Shops, schools and hospitals usually use fluorescent lights because they cost less to run as less electricity is wasted as heat.

 1 Why do fluorescent lights use less electricity than filament bulbs?

B **Filament bulb in a house.**

Some equipment transfers energy better than others. **Efficient** equipment transfers a lot of the supplied energy into the forms of energy we want. Fluorescent lights are more efficient than filament bulbs because they change more electricity into light.

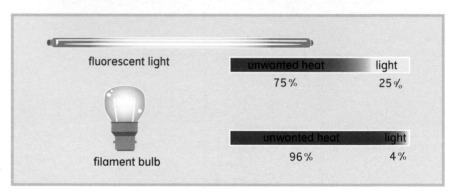

| fluorescent light | unwanted heat | light |
| | 75% | 25% |

| filament bulb | unwanted heat | light |
| | 96% | 4% |

C

E The fraction of energy *usefully* transferred is the equipment's efficiency.

The efficiency can be worked out if you know:

● how much useful energy comes out of the equipment
● how much energy is supplied to the equipment.

   efficiency of the equipment = useful energy coming out ÷ total energy supplied
   (in **joules**, **J**)   (in **joules**, **J**)

Example

100 joules of electricity are supplied to a fluorescent light.
It changes 25 joules into light energy. What is its **efficiency**?

● The useful energy coming out is 25 J of light energy.
● The energy supplied is 100 J of electricity.
● The efficiency of the fluorescent light is:
   25 J ÷ 100 J
   = 0.25

 Blackpool illuminations now use energy efficient bulbs, saving thousands of pounds each year.

A 100 W light bulb only changes 4 joules per second into light, so its efficiency is 0.04. More electricity is needed by filament bulbs to provide the same amount of light as fluorescent lights. You can find out which equipment is the most efficient by comparing the amount of energy they use to do the same job.

D *It takes less energy to boil the same amount of water in the kettle than in the saucepan. The kettle is more efficient.*

E How would you compare the efficiency of different kinds of bouncing ball? How high does each ball bounce?

2 Energy efficient bulbs change 16 joules into light energy for every 100 J supplied.

   a) How much useful energy comes out of these bulbs?

   b) How much energy was supplied to these bulbs?

   c) What is the efficiency of these bulbs?

3 Cooking a potato in the microwave oven uses one unit of electricity. Five units are needed when using an electric oven. How can you tell the microwave oven is more efficient than the oven?

! Six times more energy is needed to produce an electrical cell than the energy you can get out of it.

Summary

Efficiency is the ____ of energy supplied, which is usefully ____. We can compare efficiency two ways: by comparing the ____ needed to do the same job or by calculating the efficiency.

   amount   energy   transferred

4 Put these different types of light bulb in order with the most efficient first:

A) filament bulb, efficiency = 0.04
B) fluorescent light, efficiency = 0.25
C) energy efficient bulb, efficiency = 0.16.

5 Two cranes lift concrete blocks. The Fielding crane uses 50 litres of fuel but the Robson crane needs 40 litres to do the same amount of lifting.

   a) Explain which crane is more efficient.

   b) One crane transfers 32 kJ of energy for every 100 kJ supplied. What is its efficiency?

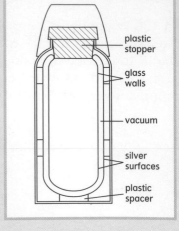

plastic stopper

glass walls

vacuum

silver surfaces

plastic spacer

1 This diagram shows part of an electricity bill

HARLOW ELECTRICITY COMPANY

Account number: 0164271
Account date: 15 July 2000

Charges for electricity

Date	Reading
11 July 2000	65128
11 April 2000	64123

Cost per Unit: 6.4 pence

a) How much did each Unit of electricity cost? (1)

b) How many Units were used in the period from April until July? (1)

c) How much did the electricity used cost? (1)

2 Here are the results of an experiment to find out how to keep a drink warm for the longest.

Exp.	Beaker	Starting temperature	Temperature after 5 min
A	no lid, no foil, no insulation	80 °C	60 °C
B	lid (no foil, no insulation)	80 °C	65 °C
C	foil (no lid, no insulation)	80 °C	68 °C
D	insulation (no lid, no foil)	80 °C	73 °C

a) List 3 things the student should have done to make sure it was a fair test. (3)

b) What type of heat transfer is she stopping in each of experiments B, C and D? (3)

c) In which experiment was the heat loss biggest? (1)

d) Looking at her results, what one thing should she do to keep a drink as warm as possible? (1)

3 This diagram shows a vacuum flask, designed to keep drinks hot.

This table compares the ways that heat is transferred in the flask. Write down what should go in boxes **a)** to **d)**. (4)

Type of heat transfer	Takes place mainly through	Can be stopped in a vacuum flask by using
conduction	**a)**	**b)**
convection	**c)**	**d)**
radiation	empty space	shiny coating on the glass

4 The table shows ways that a house may be insulated.

Where heat is lost	How heat loss can be reduced	Cost of reducing the heat loss	Amount saved each year
floor	lay carpets	£600	£150
door	draught proofing	£10	£100
roof	install fibreglass loft insulation	£350	£300

a) Explain how fibreglass insulation helps to reduce the heat loss from the roof. (1)

b) Explain which method of insulation would save you money quickest. (2)

c) Copy and complete these sentences using words from the box. You may use each word once, more than once, or not at all.

Gaps underneath doors allow heat to escape by ____. Carpets are made from good ____. They stop heat escaping by ____. (3)

cold	conductors	conduction
convection	heat	insulators	radiation

5 Look at these energy resources used to generate electricity:

> hydro-electricity   wind power   oil
>   wave power   tidal power

a) Explain why oil is the odd one out. (2)

b) Many people would like us to use renewable energy resources more often. Explain one benefit of this. (1)

c) State one reason why each of the energy resources listed below is not used more often:

   i) tidal power   ii) wave power

   iii) solar power. (3)

6 The diagram below shows a power station.

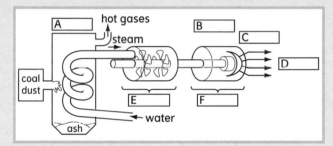

When a fuel is burned in the **furnace**, the heat released is used to change water into steam. The jet of steam forces a **turbine** to spin. The shaft of the turbine is connected to a **generator**, consisting of a **magnet** surrounded by **coils of wire**. The magnet spins. This generates **electrical energy** in the coils which can then be carried to homes and businesses.

a) Which of the bold words in the passage should go in each of the boxes A to F? (6)

b) Energy is changed into different forms in the power station. Copy and complete the table using words from the box. You may use each word once, more than once, or not at all. (4)

Part of the power station	Energy type that the part has
fuel	
heated water	
spinning turbine	
cables	

> chemical   electrical   gravitational
> potential   heat   kinetic   light   sound

7 a) Copy and complete this sentence.

A hairdryer is designed to change electrical energy into ____ energy and ____ energy. (2)

b) The power of the hairdryer is 1.6 kW. How many kilowatt-hours (kWh) of energy are transferred when the hairdryer is used for 2 hours? (2)

c) When the hairdryer is used, not all the energy is transferred usefully. What form of wasted energy is given out by the hairdryer? (1)

d) Explain where this wasted energy goes to. (1)

8 Two friends are comparing their computers. The power rating of computer A is 1.5 kW, and for computer B it is 1200 W.

a) Explain what is meant by power. (1)

b) Which computer uses more power? (2)

c) How much energy is used (in kWh) when computer A is on for 4 hours? (1)

d) Explain what energy transfer happens when computer B is switched on. Include the wasted form of energy. (2)

9 Look at this table comparing 3 fridges.

Fridge	Price	Annual running cost	Efficiency rating
A	£120	£30	0.7
B	£109	£38	0.6
C	£170	£24	0.9

a) Explain what is meant by efficiency. (1)

b) Which fridge is most efficient? (1)

c) When you buy a fridge, it is important to compare the running costs as well as the price. Calculate the running costs for each fridge over 3 years. (3)

d) Explain which fridge would be cheapest over 3 years. (2)

Voltage and current

What are voltage and current?

It is hard to imagine life without electricity. There would be no televisions, computers, telephones or even electric lights. Electricity is a form of energy called electrical energy. When we use electricity we are changing electrical energy into other types of energy, such as light, heat or sound.

 1 In a radio, what type of energy is electrical energy changed into?

Everything is made up of tiny particles. Some of these particles are called **electrons**. In certain materials, like metals, these electrons can be made to move. When electrons move we say that there is an **electrical current**. The more electrons that are flowing, the larger the current.

Current is measured in **amperes** (or **amps**). The symbol for amps is **A**.

2 Choose the correct answer. Current is measured in ...

   A) cm   **B)** amps   **C)** joules.

A

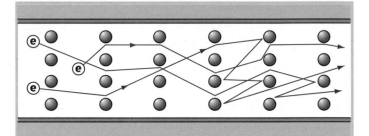
B *Electrons flow along a wire. They bump into atoms in the wire as they travel along it.*

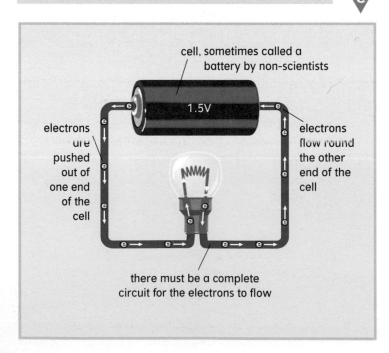
C

cell, sometimes called a battery by non-scientists

1.5V

electrons are pushed out of one end of the cell

electrons flow round the other end of the cell

there must be a complete circuit for the electrons to flow

Electrons will only flow when they are given electrical energy. The energy is often produced by an electrical cell. If the cell is in a complete circuit, the electrons are pushed out of one end of the cell, round the circuit to the other end. The size of the push is called the **potential difference** or **voltage**. The bigger the potential difference, the bigger the current that will flow around the circuit.

Voltage is measured in **volts**. The symbol for volts is **V**.

 3 What are the units for measuring potential difference?

 An electric catfish can produce a voltage of 450 V to capture its prey and defend itself.

P How could you find out the voltage of the most common cell used in portable electrical devices?

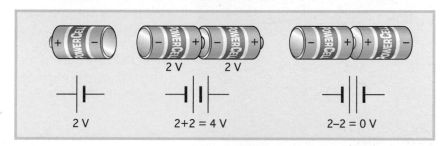

Cells in series

Different devices need different voltages to work, so more than one cell has to be used. Two or more cells used together are called a **battery**.

If two cells are connected so that they push in the same direction then their voltages are added together. If they are pushing in opposite directions then they can cancel each other out.

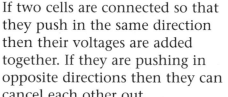

2 V

2 V   2 V

2 V   2+2 = 4 V   2−2 = 0 V

? **4** How many 1.5 V cells would be needed to run a 6 V radio?

Electrical symbols

Many electrical **components** are used in circuits. We draw **circuit diagrams** to show how the components are wired together. Some components are difficult to draw, so we use symbols.

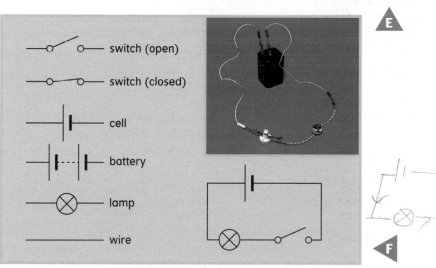

switch (open)

switch (closed)

cell

battery

lamp

wire

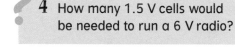

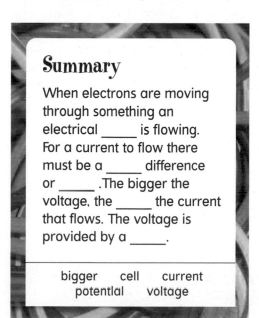

Summary

When electrons are moving through something an electrical _____ is flowing. For a current to flow there must be a _____ difference or _____ .The bigger the voltage, the _____ the current that flows. The voltage is provided by a _____.

bigger   cell   current
potentlal   voltage

?

5 Draw a circuit diagram containing two bulbs, a cell and a switch. Label each component.

6 Look at diagram G. Each cell has a voltage of 1.5 V. Work out how much voltage each combination would give you:

(a)   (b)

(c)   (d)

7 Why should cells always face the same direction in a circuit?

Measuring current and voltage

How can we measure current and voltage?

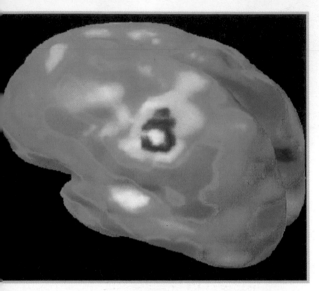

The voltages that make our brains work are tiny compared to the voltages needed to work a television.

Different devices use different voltages and currents. A computer uses a small current. An electric fire uses a large current. A torch only needs about 6 V to operate, whereas an iron needs 230 V.

We measure current with an **ammeter**. The ammeter measures the size of the current in **amps** (**A**). It is placed in a circuit as shown in diagram B.

We say that the ammeter is in series *as it is in the main part of the circuit.*

B

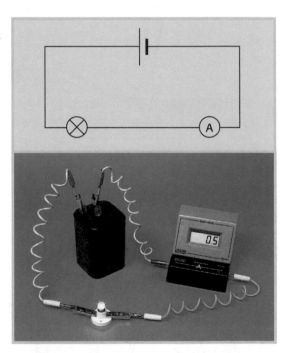

A *A brain scan. The colours show where different voltages are being produced.*

You can imagine the ammeter working like a turnstile at a football stadium. It counts the number of electrons passing through it in a certain time. The more electrons that pass through it every second, the larger the current.

1 a) What instrument is used to measure current?
   b) How is it connected in the circuit?

Voltage (or potential difference) is measured with a **voltmeter**. The voltmeter measures the size of the **voltage** in **volts** (**V**). It is placed in a circuit as shown in diagram C.

The voltmeter compares how much energy the electrons have before going into the bulb and how much they have when they leave the bulb. The more energy they lose in the bulb the bigger the voltage (or potential difference) *across* the bulb. (As voltage compares two points we talk about the voltage *across* things.)

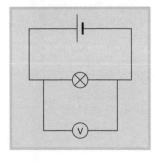

C *We say that the voltmeter is in* parallel *as it is in its own branch of the circuit.*

About one million billion electrons flow through a light bulb in one second.

2 a) What instrument is used to measure the potential difference across something?
   b) Should a voltmeter be placed in series or in parallel?

P How would you find out what happens to the current through a bulb when you change the voltage across it?

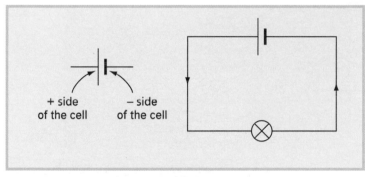

+ side of the cell   − side of the cell

E *The arrows show the current flows to the negative (−) end of the cell.*

Labelling circuit diagrams

The direction of the current flowing in a circuit needs to be labelled on the circuit diagram. We draw arrows on circuit diagrams to show the current travelling from the positive (+) end of the cell to the negative (−) end. On the cell symbol the positive end is the longer line and the negative end is the short line.

Summary

Current is measured in _____ (A) using an _____. We say the ammeter is placed in _____ because it is in the main part of the circuit. Potential difference or voltage is measured in volts (V) using a _____, which is placed across a component. We say a voltmeter is placed in _____, because it is not in the main part of the circuit.
The current in a circuit flows from the _____ end of the cell to the _____ end.

amps   ammeter   negative
parallel   positive   series
voltmeter

?

3 Draw the circuit you would make to measure:

   a) the current through a bulb
   b) the voltage across the bulb.

4 Draw and label the symbol of a cell to show the positive and negative sides.

5 Match the following statements and write out the correct sentences:

Voltage is measured *across* components	as it measures the amount of charge flowing through the component per second.
The current flowing through components can change	as it compares the energy of electrons at two points.
Current is measured *through* components	as it depends on the voltage across the component.

209

Series circuits

What happens to voltage and current in a series circuit?

All electrical circuits can be divided into two main groups:

● **series** circuits
● **parallel** circuits.

Currents in series circuits

A series circuit has only one route for the current to travel round. Diagram A shows some examples.

In each circuit the current must travel through all of the components. If a current of 3 A leaves the cell, then a current of 3 A will travel through the light bulb, wires and ammeter back to the cell.

If one bulb in the circuit breaks then there will be a gap in the circuit, so the current cannot flow. This will make the other bulb go out.

> **Rule 1:** *In a series circuit the current is the same all the way round the circuit.*

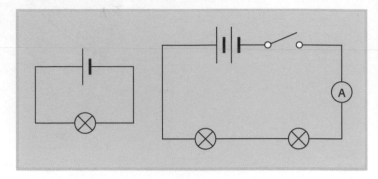

A

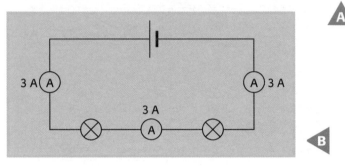

B

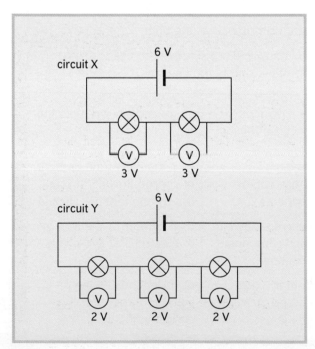

C

? 1 If the current leaving one end of a cell is 5 A, what will the current be when it is returning to the other end of the cell?

2 If there are two bulbs in a series circuit and one of them breaks, what will happen to the other bulb?

Voltage in a series circuit

The potential difference provided by the cell gives the electrons energy. The moving electrons transfer this energy to the other components in the circuit. The electrons transfer some energy to each component as they pass through, so that by the time they return to the cell they have given up all of their energy.

Look at diagram C. In circuit X, a voltage of 6 V is provided by the battery. The current delivers half of its energy to each bulb, as they are identical. Each bulb will have 3 V across it.

In circuit Y there are three identical bulbs so the 6 V will be divided equally between the three. Each bulb will have a voltage of 2 V across it.

Rule 2: In a series circuit the potential difference (voltage) is shared between the components in the circuit.

In the examples in diagram C the bulbs are identical, so the voltage is shared equally between them. In some circuits the components are different and the voltage may not be shared out equally.

? 4 Look at diagram E. Work out the correct currents and voltages where there are question marks. (All the bulbs are identical.)

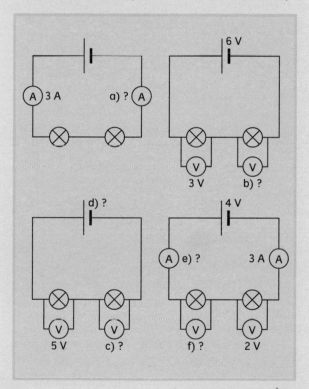

E

5 A 6 V cell operates two bulbs in a series circuit. One bulb uses twice as much energy as the other. What will the voltage be across this bulb? Choose the correct answer.

A) 2 V   **B)** 3 V   **C)** 4 V.

P How can you show that the two series circuit rules are correct?

? 3 A cell in a series circuit has a potential difference of 3 V. If there are two identical bulbs in the circuit, what is the voltage across each bulb?

D

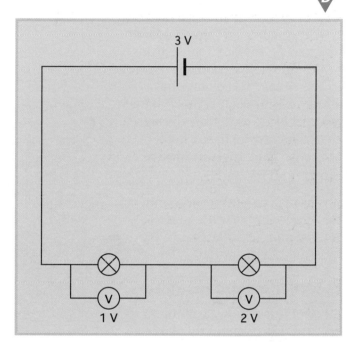

Summary

A series circuit has only one _____ for current to travel round. All the _____ must travel through all the _____ in the circuit. There are _____ series circuit rules:

Rule 1: The current is the _____ all the way round the _____.

Rule 2: The voltage is _____ between the _____ in the _____.

circuit   components   current   route
         same   shared   two

211

F4 Resistance

Why do electrons need energy to travel around a circuit?

A car needs fuel to move. A person needs food to move. The same applies to electrons. They need energy to move around a circuit.

Resistance is a way of saying how easy or difficult it is for electrons to flow through something. Look at diagram A. It is easy for electrons to flow along the connecting wires in the circuit. These wires have a **low resistance**. It is much harder for the electrons to flow through the filament wire in the bulb. This wire has a **high resistance**.

The *greater the resistance* the harder it is for the current to flow and *the more energy* that is converted to heat and light. Resistance is measured in **ohms** (Ω).

Conductors and insulators

Electrons can flow easily in **conductors**. Conductors have a low resistance. Metals are good conductors. **Insulators** have a much higher resistance. Currents cannot usually flow through insulators. Plastics are good insulators.

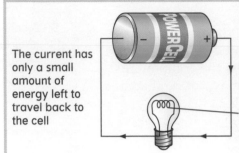

The current has only a small amount of energy left to travel back to the cell

Current leaves the cell with plenty of energy

A small amount of energy is changed to heat energy as electrons flow through the wires

The thin filament wire in the bulb is much harder for electrons to travel through. The electrons change nearly all their energy into heat and light travelling through the bulb

 B *The three-bulb circuit has more resistance so the current flowing is smaller.*

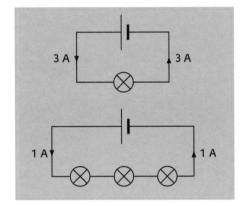

3 A   3 A

1 A   1 A

 1 Where do the electrons get their energy from in a circuit? Choose the correct answer.

A) bulb
B) wire
C) cell.

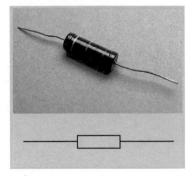

C *A resistor and its symbol.*

? 2 If a current flows through a high resistance bulb and then through a low resistance buzzer, where will the electrons give up most of their energy?

Resistors in series

Every component has a resistance. Look at diagram B. There are three times as many bulbs in the second circuit, so there is three times the resistance.

A **resistor** is a component that decreases the current in a circuit. When electricity flows through a resistor, some of the electrical energy is transferred as heat energy.

When resistors are connected in **series** their resistance can be *added* together to give the total resistance in the circuit.

The total resistance in this circuit is 6Ω. **D**

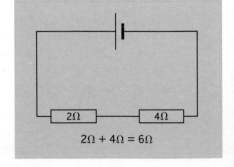

2Ω   4Ω

2Ω + 4Ω = 6Ω

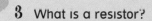

3 What is a resistor?

4 If each bulb has a resistance of 20 Ω, what is the total resistance of a series circuit with

a) 2 bulbs in it?
b) 4 bulbs in it?

E

A variable resistor is used in circuits when you want to change the resistance in the circuit easily.

P How would you find out whether different lengths of wire have different resistances?

G

Changing currents

A component with a large resistance needs a bigger voltage to push a certain current through it. If the voltage is not increased then the current will be smaller. Look at diagram F. **F**

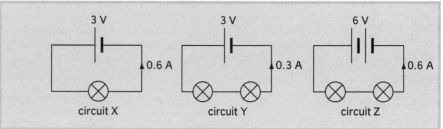

circuit X   circuit Y   circuit Z

Circuit Y has twice the resistance of circuit X. However, there is no extra voltage and so the current is halved.

Circuit Z has twice the resistance of circuit X and also has twice the voltage so the current stays the same.

E The voltage can be worked out if you know

● the current (in amps)

● the resistance (in ohms).

voltage = current × resistance
(in **volts, V**) (in **amps, A**) (in **ohms, Ω**)

If the voltage across a resistor doubles, the current through the resistor will double. If the voltage becomes three times as big, the current must be three times as big. This means the current flowing through the resistor is said to be **proportional** to the voltage across it.

5 What is the voltage across a 10 Ω resistor with a 2 A current flowing through it? Choose the correct answer.

A) 0.2 V **B)** 5 V **C)** 20 V.

Summary

Resistance is measured in _____ Every component in a circuit has resistance. The greater the resistance the _____ current that will flow if the voltage stays the same. The voltage across a resistor is _____ to the current flowing through it. _____ have higher resistances than _____.

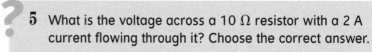

conductors   insulators   less   ohms   proportional

6 True or false?
A piece of nylon thread has a higher resistance than a piece of copper wire. Explain your answer.

Current–voltage graphs

What happens when the current in a component changes?

Some substances change resistance when they are heated or when light is hitting them. These substances can be used as detectors. A heat sensor in a burglar alarm is an example of this.

Using graphs to show resistance

We can use graphs to see how much resistance a component has.

Graph A shows what happens to the current through two resistors X and Y as the voltage across them changes. The *slope* of the line represents their *resistance*. The *steeper* the line the *lower* the resistance.

 1 Look at diagram A. Which resistor, X or Y, has the greatest resistance?

The lines on graph A are straight, which shows that the resistance of a resistor stays the same (constant). The temperature of the resistor must also stay the same.

Current–voltage graph for a light bulb

Most components do not have a constant resistance. As the current going through them increases, the resistance changes. On a graph of current against voltage, the line will not be straight. If the resistance is *decreasing* the line will become *steeper* (curve upwards). If the resistance is *increasing* the line will become *less steep* (become flatter).

The thin wire in a light bulb is called the **filament**. As the voltage and current get bigger the filament wire becomes hot. The heat energy makes the particles in the wire vibrate more, making it harder for the electrons to travel through the wire. This makes the resistance increase, so the line becomes *less steep*. The hotter the filament gets the higher its resistance.

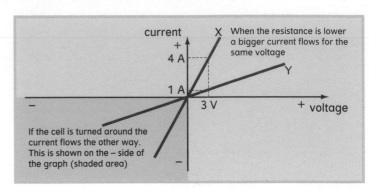

A *Resistors at constant temperature.*

B *Decreasing resistance.*   *Increasing resistance.* **C**

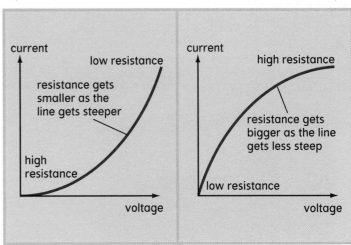

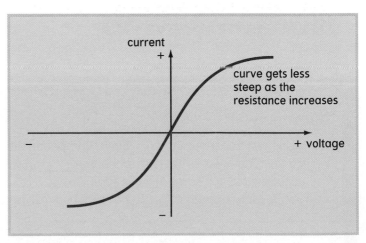

D *A current–voltage graph for a light bulb.*

Current–voltage graph for a diode

A diode is a special component that only lets current flow through it in one direction. When the current is in this direction the diode behaves just like a resistor. If the current tries to flow the other way the diode's resistance becomes extremely high so the current cannot flow.

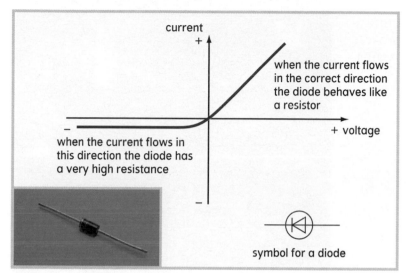

current
+

when the current flows in the correct direction the diode behaves like a resistor

+ voltage

when the current flows in this direction the diode has a very high resistance

symbol for a diode

F

Some components change resistance for other reasons.

The resistance of a light dependent resistor decreases if the amount of light shining on it increases. It is affected by the **intensity** of light.

The resistance of a thermistor decreases if it gets hotter. It is affected by temperature.

P How would you draw the current-voltage graph for a piece of wire? What circuit would you need to set up to find the measurements you need?

3 Which component normally has a constant resistance?

4 Choose the correct answer. A thermistor increases its resistance when ...

A) light gets brighter
B) temperature increases
C) temperature decreases.

5 Draw a graph of temperature against resistance for a thermistor.

2 Graph E shows the current–voltage relationship for an electric heater.

a) Is the resistance getting higher or lower as the voltage increases?

b) Up to what voltage does the resistance remain constant?

E

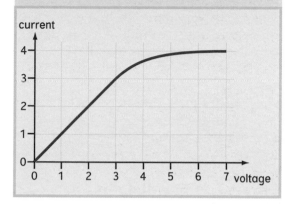

current

4
3
2
1
0

0 1 2 3 4 5 6 7 voltage

G

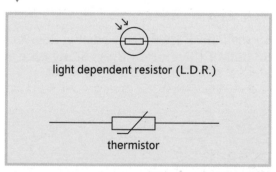

light dependent resistor (L.D.R.)

thermistor

Summary

The resistance of a component is not always _____. When the current through a bulb _____ its resistance increases. On a current-voltage graph if the line becomes less steep the resistance is going _____. Current can only flow in one direction through a _____.

constant   diode
increases   up

Parallel circuits

What happens to voltage and current in a parallel circuit?

Parallel circuits are more useful than series circuits. If there are two bulbs in a series circuit they both have to be switched on and off at the same time. This is not very useful at home. If one bulb broke, the whole house would be left in darkness! Parallel circuits solve problems like these.

1 a) Are most Christmas tree lights in series or in parallel?

b) How do you know?

Current in parallel circuits

In a parallel circuit there is more than one route for current to flow along. The current leaving the cell *splits up* and takes different routes.

In diagram B the circuit has two identical routes, so half of the current flows along each route.

2 A + 2 A = 4 A

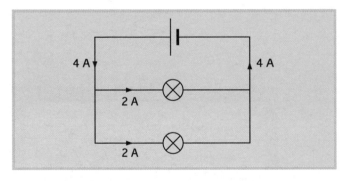

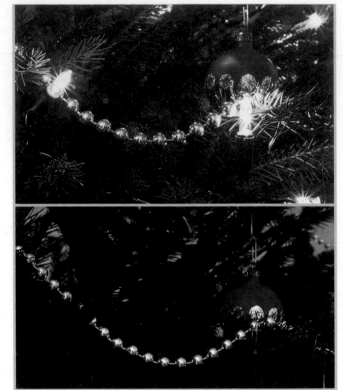

In diagram C the circuit has three identical routes so a third of the current flows along each route.

1 A + 1 A + 1 A = 3 A

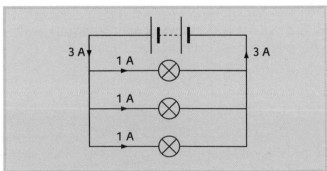

Rule 1: The current in a parallel circuit splits to go down different routes. The total current through the cell is the sum of all the separate currents.

2 A parallel circuit has two identical routes. A total current of 10 A flows through the cell. How much current will flow along each route?

How would you investigate what happens to the brightness of the bulbs as more routes are added to a circuit?

Voltage in parallel circuits

Electrons leaving the cell transfer all of their energy to the circuit before returning. It does not matter which route they take. This means that the voltage must be the same across each route.

Rule 2: *In a parallel circuit the voltage is the same across each route.*

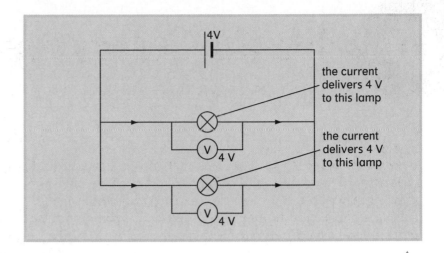

the current delivers 4 V to this lamp

the current delivers 4 V to this lamp

D

Routes with different resistances

Circuit E has two routes, with 6 V across each route. Now look at the currents. The top route has a bigger resistance, so a smaller current will flow through it than the bottom route. The current does not divide equally.

Rule 3: *In a parallel circuit the current through a route depends on the resistance. If it has a big resistance only a small current will flow along it.*

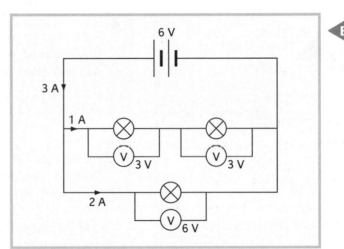

E

? 3 If a 4 V cell is placed in a parallel circuit with 3 routes, what is the voltage across each route?

Summary

Parallel circuits have more than one _____ for current to travel round. The current _____ up to travel round each route. More current will travel round the route with the _____ resistance. Each route has the same _____ across it.

lowest   route   splits   voltage

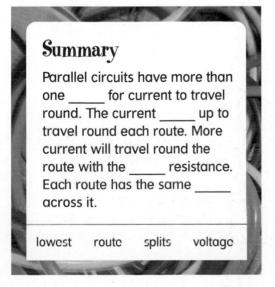

? 4 Look at diagram F. Work out the correct currents and voltages where there are question marks. (All the bulbs are identical.)

F

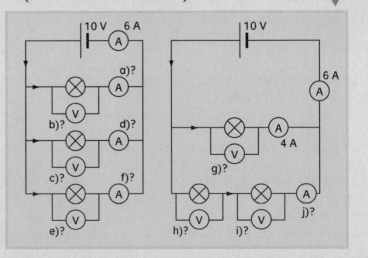

5 Design a circuit that can switch on three bulbs separately. Your circuit must also be able to switch all the bulbs off with one switch.

Static electricity

What is static electricity?

The first evidence for electrons and electric charge was discovered over 2500 years ago! When people talk about static electricity they tend to think about getting small electric shocks when they touch something. Clothes made from man-made substances, like nylon, often produce small crackling sounds when you take them off. This is caused by static electricity.

A

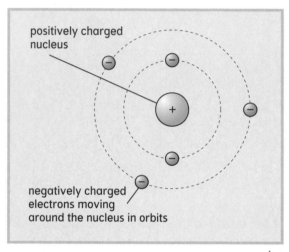

positively charged nucleus

negatively charged electrons moving around the nucleus in orbits

An atom. C

P How could you make a stream of water bend, or make your hair stand on end?

B

? 1 When you take off clothing and hear the crackle of static electricity, what do you sometimes see? Picture A may help you answer this.

Charging objects

Everything is made up of atoms. The atom has a nucleus with electrons moving around it. **Electrons** can sometimes move away from the nucleus and take their negative charge with them.

Normally, objects have equal amounts of positive and negative charge. These charges cancel each other out, so the objects are **neutral**. When two surfaces are rubbed together, the friction can rub electrons off one surface onto the other. This leaves one surface with fewer electrons and the other with extra electrons.

Electrons have a negative charge, so an object that gains electrons becomes negatively charged and an object that loses electrons becomes positively charged.

Rubbing surfaces causes friction.

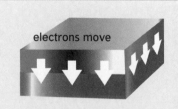

electrons move

Friction rubs electrons off one surface onto the other.

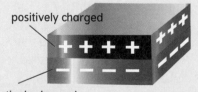

positively charged

negatively charged

Friction leaves one object with less electrons (positively charged) and the other object with extra electrons (negatively charged).

D

2 a) A balloon is rubbed on a jumper, and it becomes negatively charged. What charge is left on the jumper?

b) Have the electrons moved from the jumper onto the balloon, or from the balloon onto the jumper?

Benjamin Franklin (1706–1790) proved that lightning was caused by electricity by flying a kite in a storm. A metal key was tied to the kite to attract the charge.

Attraction and repulsion

There are two types of charge, positive and negative. Two charged objects near each other will have a force between them. This force can be either attractive or repulsive.

Two positively charged objects will **repel** each other. Two negatively charged objects will repel each other. However, a positively charged object will **attract** a negatively charged object.

> **Rule:** *Unlike* charges **attract**.
> *Like* charges **repel**.

3 Copy the sets of balls in diagram E, and draw arrows to show if the balls are attracting or repelling each other.

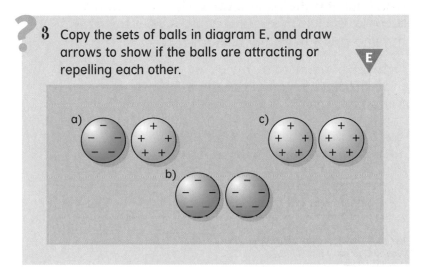

E

Attracting objects with no charge

It is possible for a charged object to attract an object with no charge.

The negative charge on the balloon repels the electrons in the wall so they move away from the surface. This leaves the surface positively charged, so the balloon sticks to the wall.

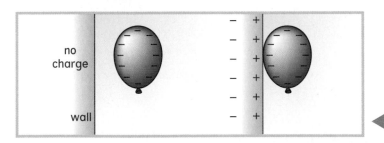

F

4 a) True or false? The nucleus of an atom repels the electrons around it.

b) Explain your choice.

5 Explain how a tiny piece of paper can be picked up using a plastic rod. (Hint: Diagram F may help you with your answer.)

Summary

There are two types of electric charge, _____ and _____. _____ have a negative charge. If an object loses electrons it becomes _____ charged. Like charges _____. Unlike charges _____. An atom is made of a _____ surrounded by electrons. The nucleus has a positive _____ that attracts the negative electrons around it. _____ can remove electrons from a surface. An object that has equal amounts of positive and negative charge is _____.

attract   charge   electrons
friction   negative
neutral   nucleus   positive
positively   repel

Using static electricity

How can static electricity be used safely?

There are about 100 lightning strikes per second on Earth! Lightning happens when clouds become charged with static electricity. The voltage this makes between the cloud and the ground can make a current flow through the air. The lightning produced can kill humans or animals that it hits.

Lightning conductors provide an easy path for charge to travel down to reach the Earth. This stops the current from damaging buildings.

 1 **a)** Why is lightning dangerous?
  b) What happens to the charge on the cloud during a lightning strike?

 Charge builds up on clouds. If the amount of charge gets large enough, the charge will jump from the cloud to the Earth as lightning.

Uses of static electricity

An **inkjet printer** uses static electricity. Droplets of ink are given a charge as they leave the nozzle. As the charged droplets pass between the plates they are attracted to the plate with the opposite charge. This means they are **deflected** (change direction). The bigger the voltage across the plates the bigger the deflection, so the position where the droplets land can be controlled. Each droplet makes a dot where it hits the paper. A group of dots in the right shape produce a letter.

 2 Why do photocopiers need to use static electricity?

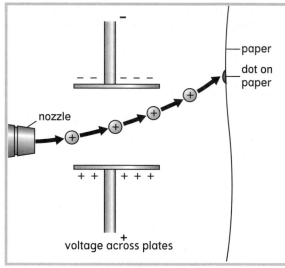

 An inkjet printer nozzle.

A A plate is electrically charged. The image to be copied is projected onto the plate. The charge leaks away where light hits the plate.

B Black powder is blown across the plate, and sticks to the charged parts of the plate.

C The plate is pressed against a sheet of paper. It is heated, the powder sticks to it, and you have a copy of the original page.

! Glowing balls, the size of footballs, have been seen during thunderstorms. They are called ball lightning. A hissing sound is heard and they melt nearby objects. Nobody knows what causes them.

 Static electricity is also used in photocopiers.

Discharging objects

When a charged object loses its charge we say it has been **discharged**.

This charged object stays charged up. The charge cannot flow anywhere.

Charge can now flow through the person to the Earth. This may give her a shock if the current is big enough.

The object is now discharged.

Charged objects can only be discharged if the charge can flow away from the object. The charge flows from a high voltage, to a place where the voltage is lower, normally the Earth.

The charge needs a conducting path to travel down. This path may be a wire, or any other conductor. For high voltages, humans or even the air can become the conducting path.

An object that is always connected to the Earth by a conductor cannot become electrically charged, because any charge on the object will flow along the conductor to the earth.

3 Find an example on these two pages, where the air is the conducting path for charge to flow through.

4 In which of the following situations is static electricity useful? Explain your answers.

a) Inkjet printers
b) Nylon clothing making sparks.
c) Photocopiers.
d) Lightning.

5 When a vehicle is filled with fuel, static electricity can build up. Explain why you think it is important for aeroplanes to be connected to Earth when they are being refuelled before a flight.

E

Summary

If electric _____ builds up on an object, the charge will want to flow to where there is _____ charge. Charge will always flow from a charged object to the _____ through the easiest path. Static electricity can be used in inkjet _____ and in _____.

charge   Earth
less   photocopiers
printers

F9 Electrolysis

What is electrolysis and how does it work?

Pure water can conduct electricity if the voltage is high enough. Sea water is a much better conductor than pure water because it has salt dissolved in it. Different liquids have different amounts of resistance.

When electricity flows through sea water some of the chemicals in the water are changed into other substances. This process is called **electrolysis**. Liquids used in electrolysis are called **electrolytes**.

Conducting electricity

Substances like common salt (sodium chloride) are made of tiny particles called **ions**. Ions are particles with a positive or negative charge. These particles can be separated.

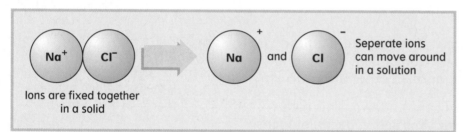

Sodium chloride (common salt, NaCl), exists as sodium ions, Na^+, and chloride ions, Cl^-.

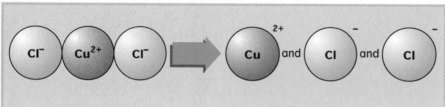

Copper chloride, $CuCl_2$, exists as copper ions, Cu^{2+}, and chloride ions, Cl^-.

Types of electrolytes

Salt water (sodium chloride solution) conducts electricity because the ions can move about in the solution. This means that charge is flowing through the liquid, so there is a current flowing. Any liquid that has ions moving in it will conduct electricity. This means that some substances that have been melted will be able to conduct electricity.

Moving ions in liquids

The ions in an electrolyte solution have charges so they can be attracted to charged objects. In electrolysis the ions are attracted to **electrodes**.

A **B**

? 1 What does sea water contain that tap water does not?

? 2 Copy table C and write the following particles into the correct columns:

sodium ion

chloride ion

electron

copper ion **C**

Positively charged	Negatively charged

3 Would a copper ion be attracted to something that is negative or positive?

When copper chloride is dissolved in water, the copper chloride splits into free copper ions and chloride ions. These ions conduct electricity.

4 Look at diagram E. If the cell is turned round:

a) will the positive electrode be on the left or the right?

b) where will the chlorine now be produced?

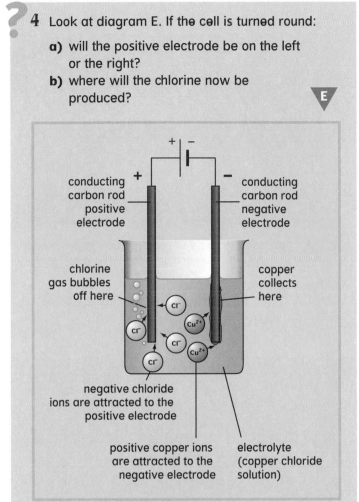

E

conducting carbon rod positive electrode

conducting carbon rod negative electrode

chlorine gas bubbles off here

copper collects here

Cl^-

Cu^{2+}

Cl^-

Cu^{2+}

Cl^-

Cl^-

negative chloride ions are attracted to the positive electrode

positive copper ions are attracted to the negative electrode

electrolyte (copper chloride solution)

How can you get pure copper from copper chloride solution?

D

copper chloride solution

5 What two elements are produced when copper chloride solution is the electrolyte in an electrolysis experiment?

6 Copy the grid below. Write the missing words into the grid and find the word in the coloured boxes.

a) The charge on a chloride ion is _____

b) The charge on a copper ion is _____

c) The liquid in electrolysis is an _____.

d) When charge moves a _____ is flowing.

e) Sodium chloride is common _____

f) Electrodes can be made from _____

g) Electrolytes can _____ current.

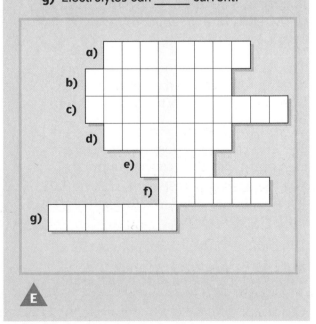

a)

b)

c)

d)

e)

f)

g)

E

Summary

Some substances in a solution split up into electrically charged _____. The _____ ions can be attracted to a positive electrode. The positive ions can be attracted to a negative _____. The electrodes are often made from _____ which conducts electricity.

carbon   electrode   ions   negative

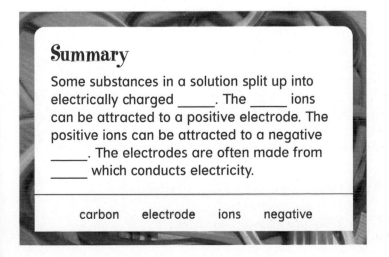

Magnetism

What is magnetism and how is it used?

You can't touch or see magnetic fields, but they can produce forces that make motors turn, or generate electricity. Magnetic substances were found naturally, hundreds of years ago, in rocks. Humans first used these magnetic materials and the Earth's magnetism for navigation.

 1 Give two uses for magnetism.

Magnetic materials

A **magnet** is something that can attract a **magnetic material**. There are only three natural magnetic elements: iron, nickel, and cobalt. Other magnetic materials can be made by mixing these elements with other materials. Steel is made from iron that has been mixed with carbon, so steel is also a magnetic material.

Only magnetic materials can be made into magnets.

Magnetic fields and poles

Magnets have **magnetic fields** around them. This is the space around the magnet where it can attract magnetic materials. The magnetic field is strongest at the ends of a magnet. These are called the **poles** of the magnet.

The Earth has a magnetic field. If a magnet can move, it will turn around until it is lined up with the north-south direction of the Earth.

The end of a freely suspended magnet that points north is called the **north-seeking pole** of the magnet. The end of the magnet pointing south is called the **south-seeking pole**.

The name north-seeking pole is often shortened to **north pole**. South-seeking pole is often shortened to **south pole**.

A *The needle can turn freely. The red end points north. It is the north–seeking pole of the compass 'needle.'*

 2 What three magnetic elements are found naturally?

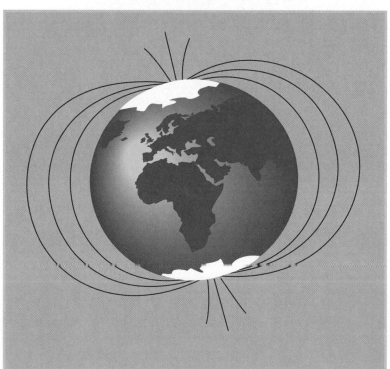

B

 3 a) Where is the magnetic field strongest on a bar magnet?
b) What are these areas called?

The rules of magnetism

A magnet can attract *or* repel another magnet, but can *only* attract a magnetic material. You can only prove something is a magnet by seeing if it will repel another magnet.

4 Two magnets are placed near each other, with their north poles facing. Will they attract or repel?

5 If two magnets are attracting each other and one of them is turned round, what will they do now?

Drawing magnetic fields

Magnetic fields can be drawn with arrows, which always point from the north-seeking pole towards the south-seeking pole.

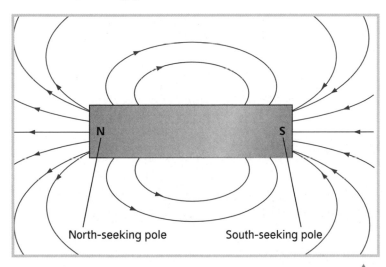

North-seeking pole   South-seeking pole

D

Summary

The magnetic field around a magnet is strongest at its _____. Like poles will _____ each other. Unlike poles will _____ each other. The end of the magnet which points to the Earth's North Pole is called the _____-_____ _____. When drawing magnetic fields the arrows on the field lines always point from the _____-_____ pole to the _____-_____ pole.

attract   north-seeking   north-seeking pole
    poles   repel   south-seeking

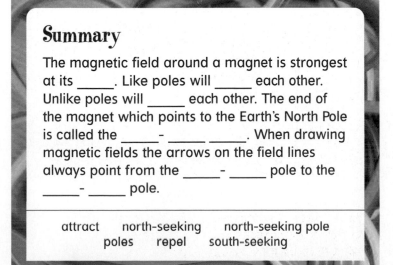

Rules:

Unlike poles will **attract** each other.

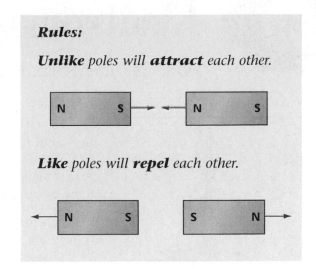

Like poles will **repel** each other.

C

P How can you use iron filings or a compass to find the shape of a magnetic field?

E

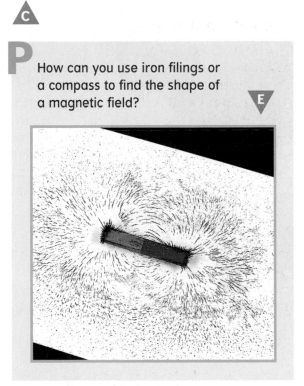

6 Which three of the following can be used to find the shape of a magnetic field?

copper filings
iron filings
aluminium filings
steel filings
nickel filings.

7 Write down how you would prove to a friend that an object is a magnet.

Electromagnetism

How can you make a magnet using electricity?

Electricity can be used to make magnets that can be switched on and off. A magnet made by electricity is called an **electromagnet**. Electromagnets are called **temporary magnets** because their magnetism can be switched on and off. This is different from **permanent magnets** which always stay magnetic.

Electromagnets are used in industry. For example old cars are picked up by electromagnets to move them around in scrap yards. To put the car down again the electromagnet is switched off.

 1 What can an electromagnet do that a permanent magnet cannot?

Magnetic fields around currents

Anything with a current flowing through it has a magnetic field around it. If you plug in any electrical device and switch it on, there will be a magnetic field around the cable. There are also magnetic fields around the large electricity cables carried by pylons that bring electricity to your home. You are surrounded by magnetic fields, but many are so weak that they are hard to detect.

 2 What happens to the magnetic field around an electromagnet if the current stops flowing through the electromagnet?

To make the magnetic field stronger, loops of wire are used. These are normally in the shape of a **coil**.

A

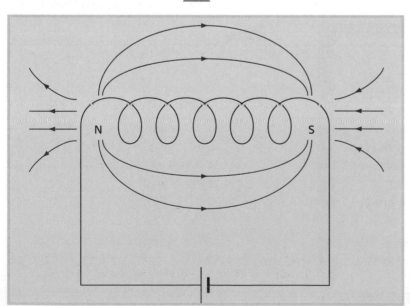

The coil acts just like a bar magnet, but it can be switched on and off as the current is switched on and off. The ends of the coil act like poles. If the current flows the opposite way around the circuit, the poles swap around.

 3 Re-draw diagram B with the cell turned the other way around. (Be careful which way you draw the arrows.)

B

Increasing the strength of an electromagnet

Sometimes an electromagnet has to have a strong magnetic field.

There are three ways to increase the strength of an electromagnet:

- By placing an **iron core** inside the coil.

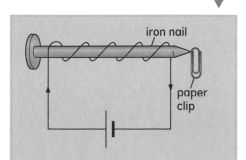

- By increasing the number of turns on the coil.

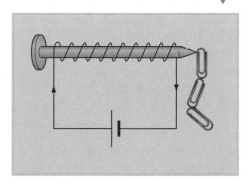

- By increasing the size of the current through the coil.

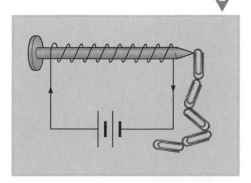

The coil needs to be **insulated** so no current flows into the iron core.

How would you investigate what happens to the strength of the electromagnet if you increase the current?

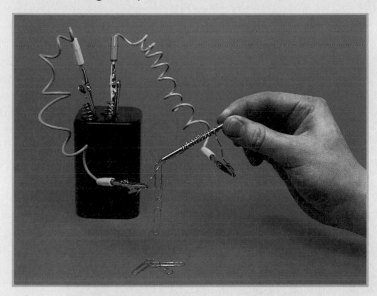

F *The stronger the electromagnet, the more paper clips it can pick up.*

4 What are the three ways to increase the strength of an electromagnet?

5 Which of the following cannot be used to insulate the coil of an electromagnet? Choose the correct answer.

A) plastic   **B)** copper   **C)** rubber.

6 Explain why the coil of an electromagnet needs to be made from wire with a plastic coating.

Summary

Anything with a current flowing through it has a _____ field around it. Electromagnets can be switched on and off. They are _____ magnets. Electromagnets are used in scrap yards to pick up _____ car bodies. The strength of an electromagnet can be _____ by putting an _____ _____ inside the coil, by increasing the _____ _____ _____ or by increasing the _____.

current	increased	iron core	magnetic	number of
	turns	steel	temporary	

The motor effect

How can the field around a current be used to turn motors?

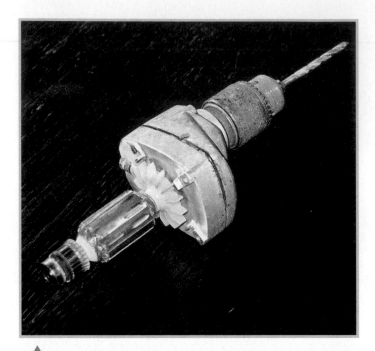

A

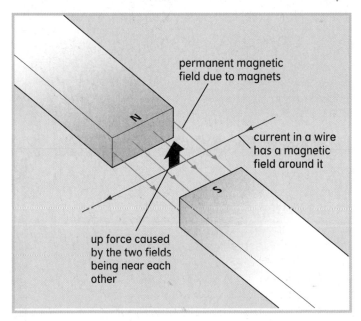

permanent magnetic field due to magnets

N

current in a wire has a magnetic field around it

S

up force caused by the two fields being near each other

P How would you make a piece of kitchen foil move without touching it?

After scientists had discovered that currents had magnetic fields around them, they invented devices that made use of this new discovery. One of the most useful of these devices is the electric motor. Many of the objects that we use at home have motors inside, for example, hair dryers and compact disc players.

? **1** Write down the name of another device in the home that uses an electric motor.

When two magnets are near to each other they will produce a force between them. This is because the two magnetic fields affect each other.

Anything with a current flowing through it has a magnetic field around it. If this magnetic field is placed near another magnetic field, it is similar to putting two magnets close together. The two fields affect each other and produce a force. This is called the **motor effect**.

In diagram B the wire will be forced to move upwards. The force never pushes the wire towards the magnet. The force produced is always at right angles to the wire.

B

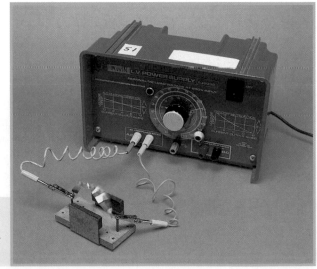

C

Changing the direction of the force

If you change the direction of the current or the magnetic field, the direction of the force changes.

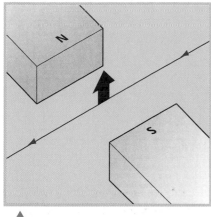

D

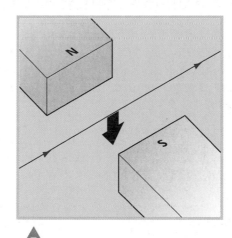

E

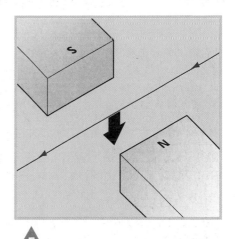

F

The force in diagram D is upwards. If the current flows the opposite way along the wire, as in diagram E, the force will be in the opposite direction. The force on the wire is now down. Look at diagram F. If the magnetic field is the opposite way round the force will also be in the opposite direction.

Changing the size of the force

Motors often need to produce large forces. There are two main ways to increase the size of the force:

- Increase the strength of the magnetic field (see page 227).

- Increase the size of the current in the wire which is moving.

Summary

When a wire with a _____ flowing through it is placed in a _____ field, a force is produced. This is called the _____ effect. The size of the force will get bigger if the current is _____ or the magnetic field becomes _____.

---

current   increased   magnetic
motor   stronger

 2 In diagram G the current and the magnetic field are both changed to the opposite directions. What direction will the new force be in?

G

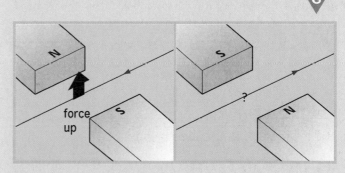

force up

3 Many electric motors use electromagnets instead of permanent magnets. How can the magnetic field be made stronger using an electromagnet?

4 Which of the following statements about the motor effect is true? Choose the correct answer.

   A) Using a bigger current will make the force smaller.

   B) Changing the current direction will not change the direction of the force.

   C) Changing the direction of the magnetic field will change the direction of the force.

5 There are two magnetic fields in diagram B. Where does each one come from?

Uses of electromagnetism

How do we use electromagnets?

When you listen to a CD player, an electric motor makes the CD go round. An electromagnet inside the loudspeakers makes the loudspeakers produce sound. These are just two uses of electromagnets.

Relays

Relays are used in circuits as safety switches. When someone turns the key to start a car, a **relay** is used to switch the current in the car on safely.

When a **small** current is switched on in the relay circuit the electromagnet is magnetised and attracts the iron **armature** towards it. The armature turns on a pivot and pushes the contacts together. This switches on a **large** current to work the motor in the main circuit. This is a safe way to switch on large currents.

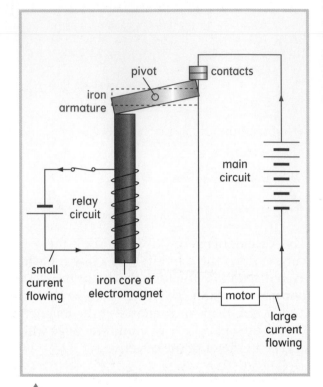

A

1 Why is the armature made of iron?

2 Look at diagram A. Why is using a relay safer than pressing a switch in the main circuit?

Loudspeakers

Loudspeakers also use electromagnets.

Look at diagram B. The current from the CD player is going backwards and forwards. When a current flows one way through the wire coil it makes the coil move to the left. When it flows the other way it moves to the right. The cone moves backwards and forwards. These vibrations make the sound waves.

3 What would happen to the movement of the cone if:

a) the electric current changes direction more quickly?

b) there is a bigger electric current?

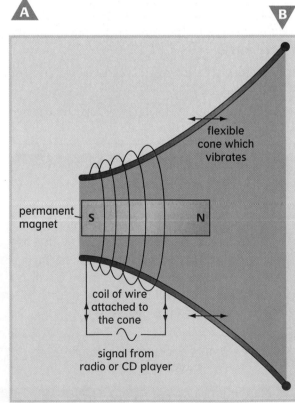

B

Electric motors

Look at diagram C. The current in the wire loop has its own magnetic field, which is inside the magnetic field of the permanent magnets. This produces a force. On one side of the coil the current is flowing from left to right and the force is upwards. Along the other side of the coil the current flows the opposite way so the force is downwards. This makes the loop turn.

There are three ways to make a motor produce more force:

- Wind more coils onto the loop.
- Use stronger magnets.
- Increase the current in the wire loop.

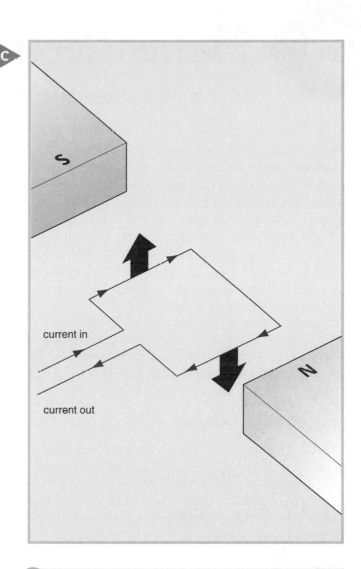

C

current in

current out

S

N

4 In picture D, the motor turns clockwise. Which way would it turn if the north and south poles of the permanent magnets were swapped around?

D

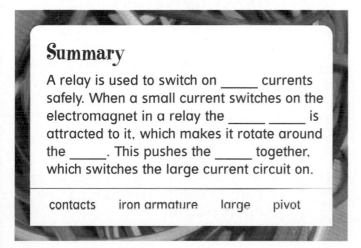

5 Copy table E and tick the statements that will make the motor spin faster.

E

a) Use weaker magnets in the motor.	
b) Wind more coils onto the loop.	
c) Use a smaller current in the coil.	
d) Use a higher voltage cell to power the motor.	

6 a) There are two magnetic fields in a loudspeaker. Where does each one come from?

b) What do two magnetic fields produce when they are near each other?

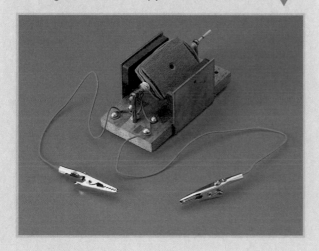

Summary

A relay is used to switch on _____ currents safely. When a small current switches on the electromagnet in a relay the _____ _____ is attracted to it, which makes it rotate around the _____. This pushes the _____ together, which switches the large current circuit on.

contacts   iron armature   large   pivot

F14 Mains electricity

How is using mains electricity different from using batteries?

People know that electricity is dangerous but because they use it every day they forget how easy it is to have accidents. An electric shock can burn you or even kill you. You can get an electric shock when your body becomes the easiest path for a current to travel down into the Earth. Faulty plugs or appliances can give you an electric shock. Water can also conduct electricity at high voltage, so you must never use electrical equipment with wet hands.

 1 Why should you dry your hands before using electrical equipment in the kitchen?

The UK mains supply

Electricity companies supply houses and flats with electricity at 230 V. The size of the current that flows depends on the resistance of the device being used. A radio with a high resistance will have a small current flowing through it. A device with a lower resistance, such as an electric shower, will have a much higher current flowing through it.

Humans have quite a high resistance, but a voltage of 230 V will make a current of around 2 A flow through a human body. This is a large enough current to cause severe injury or death.

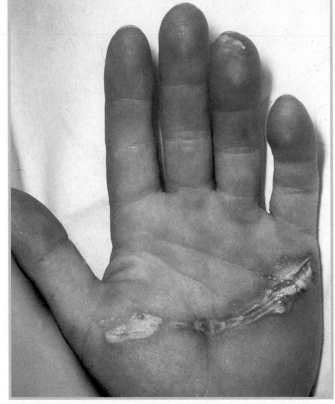

A

2 What is the voltage of the UK mains supply?

3 The light switch in a bathroom is usually a pull cord rather than a wall switch. Explain why.

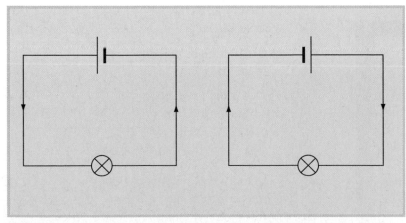

B Direct current always flows one way around the circuit. You can only change it by taking out the cell and replacing it the other way round.

Direct current and alternating current

The electricity you get from a cell is not the same as the electricity from the mains supply. The voltage of the mains supply is much higher, but also the current from the mains supply is constantly changing directions.

A cell supplies a current that always flows in one direction. This is called **direct current**.

232

The mains electricity supplies current that is constantly changing direction. In fact it changes direction 100 times per second. (It has a frequency of 50 hertz or 50 cycles per second.) One cycle is one change in direction followed by a second change in direction. This would be like changing the direction of a battery 100 times in a second to make the current go backwards and forwards. This is called **alternating current**.

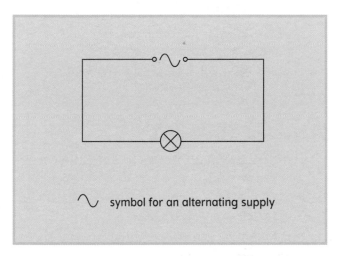

~ symbol for an alternating supply

C *The direction of the current in an alternating current supply is constantly changing. An alternating current supply has a different symbol.*

Summary

_____ current always flows in one direction. Alternating current is constantly _____ direction. Batteries supply _____ currents. Mains electricity is always an _____ current.

alternating   changing   direct

? 4 Write down whether each of the following appliances uses direct current or alternating current.

   a) battery operated calculator
   b) washing machine
   c) television
   d) mobile phone.

5 Why don't you get an electric shock from a 1.5 V battery?

6 Find out the mains supply voltage in another country.

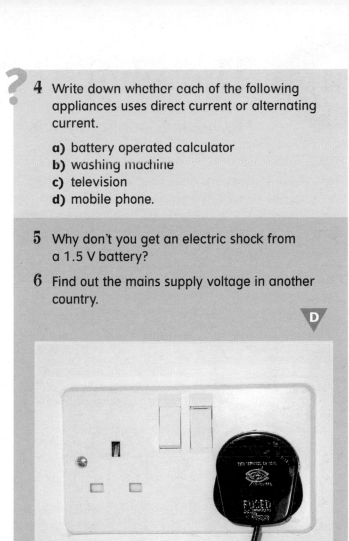

A plug socket in the UK …

… and one in Portugal.

The plug

How should a plug be wired?

Electric appliances are now sold with plugs fitted, and some of these have plastic moulded round them so they cannot be removed. This was done to reduce the number of accidents in the home, because many people do not know how to wire a plug properly.

The plug and the cable

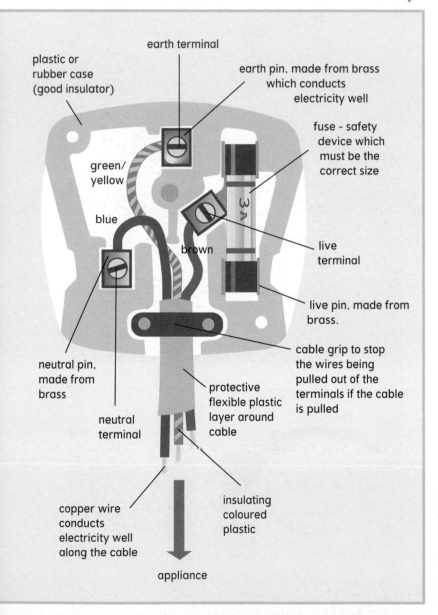

plastic or rubber case (good insulator)

earth terminal

earth pin, made from brass which conducts electricity well

fuse - safety device which must be the correct size

green/yellow

blue

brown

live terminal

live pin, made from brass.

neutral pin, made from brass

neutral terminal

cable grip to stop the wires being pulled out of the terminals if the cable is pulled

protective flexible plastic layer around cable

copper wire conducts electricity well along the cable

insulating coloured plastic

appliance

The cable grip stops the wires being pulled out of the terminals when this happens! **B**

? 1 How many wires are there inside a cable?

2 What are the colours used on the insulating cover of each wire?

3 Which terminal on the plug is connected to the fuse?

4 a) What is the cable grip for?
   b) Write down an example of a situation where the cable grip would be needed.

How to wire a plug

Here are the main things to remember when you are wiring a plug:

- Make sure you connect the coloured wires to the correct **terminals**.
- Make sure the copper wires are tightly screwed into the terminals.
- Make sure the wires are not sticking up in the plug. The top must fit on easily.
- Have the outer plastic coating of the cable right under the **cable grip.**
- Make sure the cable grip is screwed down tightly.
- Make sure the fuse is the correct size for the appliance (see page 236).
- Don't have strands of copper wire that aren't screwed into a terminal.

? 5 Which of the tools in the list below would you need to wire a plug correctly?

screw driver	spanner
hammer	sand paper
wire stripper	wire cutters

The three wires

The **live** and **neutral** wires carry the electricity for the appliance. The third **earth** wire is for safety and usually no current travels down it. Appliances that have a metal case have the earth wire connected to the case. If the case accidentally becomes charged, there is an easy path for the electricity to travel down to the ground, instead of it travelling through a person who touches the case.

The earth wire has a low resistance. If a fault in the appliance connects the case to the live wire, a large current flows which breaks the fuse in the plug. This stops the current flowing. It also means that someone touching the case will not get an electric shock.

Summary

There are _____ terminals on a plug: the neutral, _____ and _____. The blue wire is connected to the _____ terminal, the _____ and _____ wire is connected to the earth terminal and the brown wire is connected to the live terminal, which is next to the _____.

earth	fuse	green	live
neutral	three	yellow	

P How do you wire a plug correctly

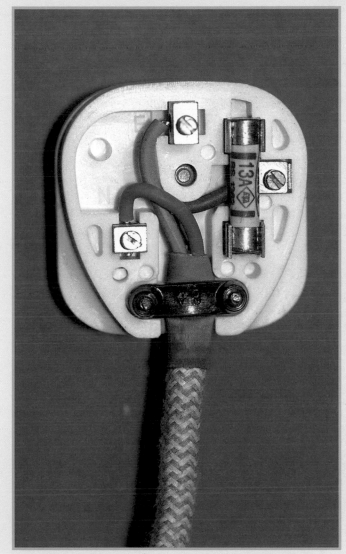

? 6 Write down three mistakes that could be made when wiring a plug.

7 True or false? Some appliances do not need an earth wire, so they have only two wires leading to the plug. Explain your answer.

! There are three holes in the electric wall sockets. If you poke anything into just one hole you can be electrocuted, it doesn't take two, as many people think!

Fuses and circuit breakers

How do fuses and circuit breakers work?

Imagine buying a new stereo system or a computer. You plug it into the mains supply and it works perfectly. Suddenly, because of a fault, there is a surge of current from the mains which flows into the new equipment, melts the wires and components inside it and breaks it.

This could easily happen. The only way to prevent this is to use a safety device that stops the current flowing if the current gets too big. These safety devices also protect people from getting severe electrical shocks.

There are two safety devices that are used to stop the current flowing when it becomes too large:

- a **fuse** (found in plugs)
- a **circuit breaker** (made from an electromagnet).

Fuse

A fuse is a very thin piece of wire held in a glass container. It is fitted **in series** with the live wire so that the electric current flows through it. If the current increases, the wire gets hotter. If the current becomes too big the thin wire melts and breaks. This puts a gap into the circuit so the current stops flowing.

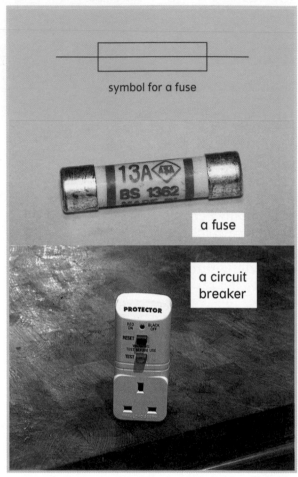

symbol for a fuse

a fuse

a circuit breaker

B

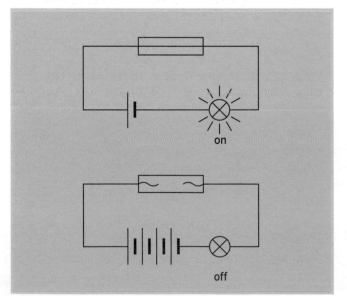

on

off

The *thicker* the fuse wire the *higher* the current needed to melt it. The **fuse rating** (size) must be correct for each appliance. If the appliance only needs 4·5 A to run then the fuse needs to melt at a current just above 4·5 A. A fuse is **overloaded** when the current becomes too large and melts it.

?

1 How does a fuse work?

2 **a)** If a heater needs 12 A to run properly, what size fuse would it need?
   Choose the correct answer.
   A) 5 A   **B)** 13 A   **C)** 25 A.
   b) Explain your choice.

Circuit breaker

A **circuit breaker** also stops current flowing if the current gets too large.

The larger current attracts the **iron armature** on the left. The end of the armature moves down so the **contacts** are separated. The current can no longer flow.

When the reset button is pressed it pushes the armature back into its original position, so the current can flow again. This should only be done after the fault has been found and corrected.

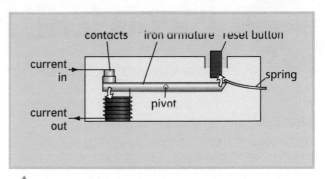

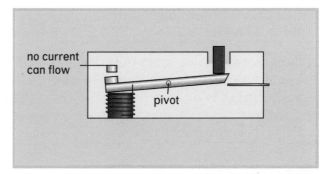

3 a) What happens to the coil when a current flows through it?
b) When the current gets bigger, why does the armature rotate?
c) Why does this stop the current flowing?
d) What does the reset button do?

Advantages and disadvantages

Fuses are cheaper to use than circuit breakers. Circuit breakers are more sensitive to increases in current and will break a circuit faster than a fuse. Therefore circuit breakers give greater protection to equipment. An electrician or electrical engineer has to decide which is the best to use in different situations.

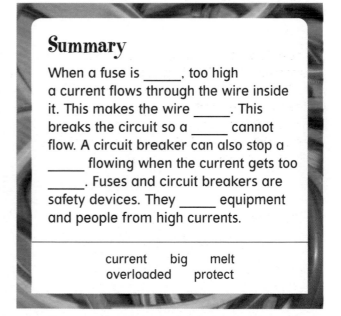

Summary

When a fuse is _____, too high a current flows through the wire inside it. This makes the wire _____. This breaks the circuit so a _____ cannot flow. A circuit breaker can also stop a _____ flowing when the current gets too _____. Fuses and circuit breakers are safety devices. They _____ equipment and people from high currents.

current   big   melt
overloaded   protect

4 a) What advantage do fuses have over circuit breakers?
b) What two advantages do circuit breakers have over fuses?

5 Match the parts of a circuit breaker to the correct descriptions.
Write out the correct sentences.

A) The reset button	1) becomes an electromagnet when a current flows.
B) The coil	2) gets attracted to the coil when a current flows.
C) The iron armature	3) pushes the armature back in place so a current can flow again.

Electrical Power

How do we work out how much energy appliances use?

The words power and powerful have many different meanings. A person is said to be powerful if they can lift heavy objects. A car is powerful if it can do 0–100 kilometres per hour in less than 5 seconds, and so on. In electricity, **power** is a measure of how quickly something transfers energy.

Power is measured in **watts (W)**. This is the amount of energy transferred in one second. Energy is measured in **joules (J)**, and so:

1 watt = 1 joule of energy transferred
        per second
1 W   = (1 J/s)

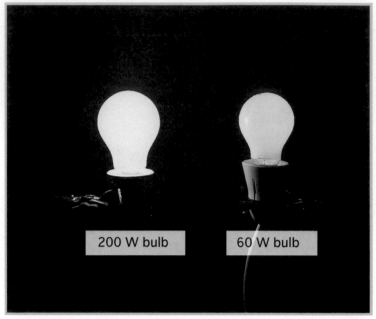

| 200 W bulb | 60 W bulb |

A

?
1 What is the unit of power?

2 How many times more powerful is a 240 watt light bulb than a 60 watt light bulb?

A 120 watt light bulb is twice as powerful as a 60 watt light bulb, as it produces about twice as much visible light energy per second. This is why it looks brighter.

As many electrical appliances convert large amounts of energy per second, we often have to work in kilowatts (kW), where 1 kW = 1000 W.

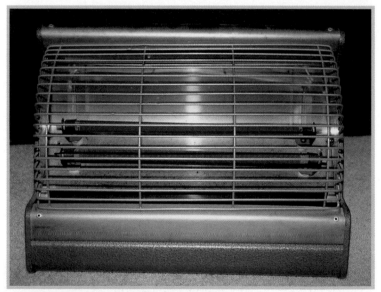

E **Calculating power**

The power of an electrical appliance can be worked out if you know:

 B *This electric fire is using 1 kW of energy. When the second bar is switched on it uses 2 kW.*

- the voltage (in volts)
- the current (in amps)

    power   =   **voltage** × **current**
(in **watts, W**)   (in **volts, V**)   (in **amps, A**)

?
3 a) How many watts of power are there in 5 kW?

    b) How many watts of power are there in 0.5 kW?

Appliances that have high power ratings convert more energy per second than appliances with low power ratings. Electrical appliances that are used to heat things generally have higher power ratings than other appliances, like televisions or radios.

Examples

An electric motor needs a voltage of 4 V to operate normally and a current of 0.1 A. What is the power of the motor?

power = voltage x current

  = 4 x 0.1

  = 0.4 W

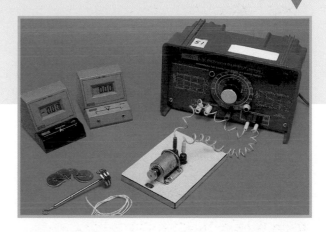

C

An electric shower is supplied by the mains supply, at a voltage of 230 V. If a current of 20 A flows through it when it is working, how powerful is the shower? Put your answer in kW.

power = voltage x current

  = 230 x 20

  = 4600 W

  = 4.6 kW

P How can you find the power of an electric motor when it is lifting different loads? You may need to use a voltmeter and an ammeter. **D** ▽

4 a) Copy out the equation for power.

  b) Copy out table E. Use the power equation to complete the table:

Power (watts)	Voltage (volts)	Current (amps)
	10	10
	4	5
	4	10
	20	3
	20	2.5

E

5 An electric shower is supplied by the mains at 230 V and has a current of 25 A flowing through it. What is the power of the shower? Choose the correct answer.

  A) 575 W   **B)** 5.75 kW   **C)** 57.5 kW.

6 Which will have a higher power rating, a kettle or a radio? Explain your answer.

Summary

Power is the amount of _____ converted per _____. _____ is measured in _____ (W). 1 watt = 1 _____ of energy transferred per second. 1 kilowatt = _____ W. The _____ of an electrical appliance is calculated using the equation

power   =   _____ × current
(in _____)   (in V)   (in _____)

| 1000 | A | energy | joule | power |
| second | voltage | W | watts | |

Generating electricity

How is electricity generated?

We use **electricity** for many things. We would find it hard to manage without it. Scientists are constantly looking for new, cheaper ways to produce electricity. Their methods use magnetism to make electric currents.

There are two ways to produce a current:

Method 1: Move a magnet in a coil of wire

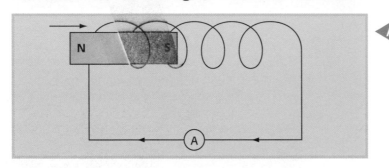

A

As the magnet is moved into the coil of wire a voltage is produced. If there is a *complete circuit* a current will flow in the wire. We say that current is **induced** in the wire.

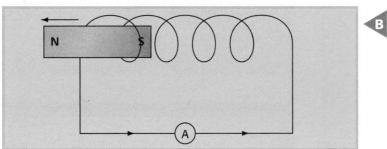

B

If the magnet moves out of the coil, the current is induced in the opposite direction.

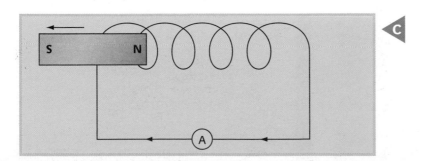

C

If the poles of the magnet are reversed, the current is induced in the opposite direction.

! Michael Faraday invented the first electricity generator at the beginning of the 19th century.

P How can you induce a current in a wire? How can you make the current bigger?

D *These Freeplay torches have small generators inside them. Just 60 turns of the handle will produce enough electricity to keep the bulb lit for 10 minutes.*

?
1 What is an induced current?
2 There are two ways to change the direction of the induced current. What are they?

Method 2: Move a wire through a magnetic field

If a wire moves through a magnetic field a voltage is induced. If there is a complete circuit a current can flow.

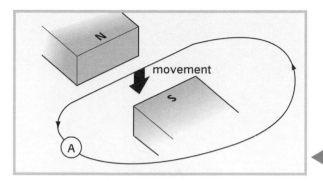

If the wire moves in the opposite direction the current will flow in the opposite direction.

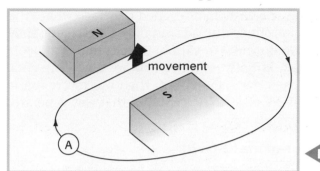

Electricity generators

Electricity generators use a very long piece of wire wound into a coil. Electricity can be generated by rotating the coil of wire in a magnetic field or by rotating a magnet inside the coil.

Changing the size of the induced voltage and current

There are four ways to increase the size of the induced voltage and current:

- Move the coil or magnet faster.
- Use a stronger magnetic field.
- Wrap more turns (loops) on the coil.
- Make the area of the coil bigger.

3 What will happen to the current if the wire moves faster through the magnetic field?

4 How could you change the direction of the current?

5 Match the sentence halves together to show the four ways to increase the induced voltage when generating electricity. Write out the correct sentences.

Make the area of the coil	greater.
Turn the coil	stronger.
Make the number of turns on the coil	faster.
Make the magnetic field	larger.

6 A bicycle dynamo works like a generator. It provides the electricity needed to run the bicycle lights. Explain why this is not useful when the cyclist stops at traffic lights.

Summary

There are _____ main ways to induce a voltage across a conductor. If there is a _____ circuit a current can flow. Method 1 is to move a _____ in a coil of wire. Method 2 is to move a _____ through a _____ field. If the direction of the movement is _____ the voltage will be reversed and the _____ will flow in the opposite direction.

complete   current   magnet   magnetic
reversed   two   wire

Transformers

What is a transformer?

When scientists had reached the stage where they could produce a steady supply of electricity, they had to find out how to supply the electricity to people's homes. Initially, families who were wealthy enough had their own generators. Then ways were found to distribute electricity to homes that was generated at power stations.

The National Grid

Power stations produce electricity, which is **transmitted** to where it is needed along cables called **power lines**. Many of these can be seen supported by tall **pylons**. This network of power lines is called the **National Grid**. It connects almost every home in the UK to an electricity supply.

A *Early generators were unreliable and very noisy.*

? **1** What voltage is the electricity in the power lines?

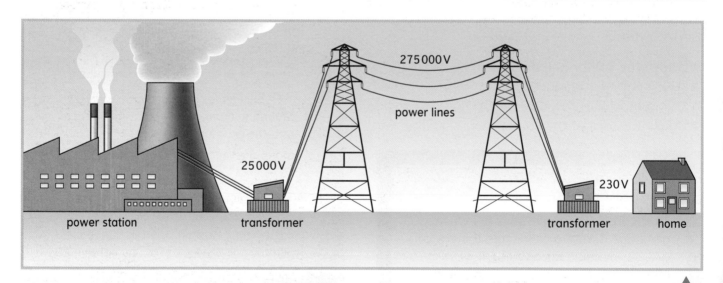

275 000 V

power lines

25 000 V

230 V

power station   transformer   transformer   home

B

Electricity has to flow along many miles of wire before it reaches people's homes. The current heats up the wire, so some of the electrical energy is wasted as heat as it travels. If the electricity is transmitted at high voltages and low currents, less energy is wasted. This makes the National Grid more **efficient**.

It would be far too dangerous to have the same high voltages in the home so the voltage is reduced to 230 V.

? **2** **a)** What would happen to the electric currents in the home if the voltage from the National Grid increased from 230 V to 275 000 V?
b) What could this size of current do to the wires in electrical equipment?

Transformers

Transformers are used to increase and decrease the voltage.

A **step-up transformer** increases the voltage.

A **step-down transformer** decreases the voltage.

Transformers are used to step-up the voltage from the power station to the power lines, and then local transformers step-down the voltage from the power lines to factories and homes. Transformers only work with alternating currents and voltages, which is why the mains supply is an alternating supply.

3 Why are the transformers near people's homes step-down transformers?

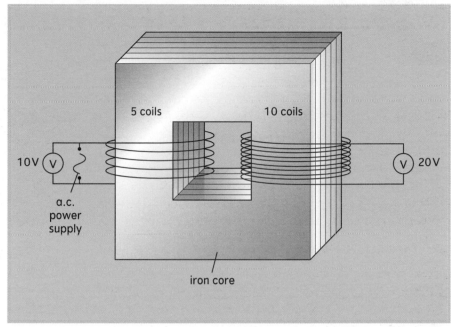

5 coils   10 coils

10V (V)   (V) 20V

a.c. power supply

iron core

A step-up transformer.

4 Write down whether a step-up or a step-down transformer is used in each of the following cases.

a) At a power station to make the voltage higher before it goes along the power lines.

b) To reduce the voltage before it reaches homes.

c) To change the mains supply to a smaller voltage for a model racing car.

Summary

Electricity is transmitted at very _____ voltages to reduce the energy wasted. These voltages are too _____ to use in the home. _____ are used to step-down the voltage before it reaches houses. The National Grid uses _____ _____ to connect almost every home in the UK to an electricity supplier.

dangerous   high   power lines
transformers

5 Explain why it is dangerous to fly kites near power lines.

Further questions

1 Name the electrical symbols shown in this table. (8)

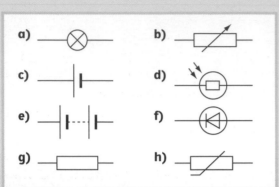

a) b) c) d) e) f) g) h)

2 Circuits A to D are made with identical cells and bulbs.

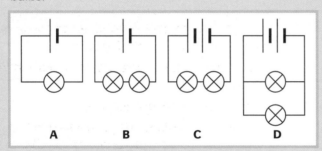

A   B   C   D

a) Which circuit has
  i) the brightest bulb(s)
  ii) the dimmest bulb(s)? (2)

b) Which two circuits have bulbs with equal brightness? (1)

3 a) What kind of charge will attract a negative charge? (1)

b) This diagram shows two charged spheres suspended close to each other. What is the charge on X? (1)

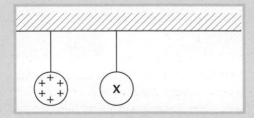

c) Describe two uses of static electricity. (2)

4 The bulbs in this circuit are identical.

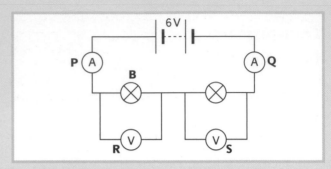

6 V
P A   B   A Q
R V   V S

a) Ammeter P reads 0.5 A. What is the reading on ammeter Q? (1)

b) The battery supplies a voltage of 6 V to the circuit. What is the reading on
  i) voltmeter R
  ii) voltmeter S? (2)

c) Each bulb in the circuit has a resistance of 6 Ω (ohms).
  i) What is the total resistance of the two bulbs in the circuit?
  ii) A current of 0.2 A now flows through one of the bulbs. Calculate the voltage across the bulb. (2)

d) Bulb B is unscrewed. What are the new readings on each of the meters, P and Q? (2)

5 This graph shows how the current in two components, A and B, changes as the voltage across them changes.

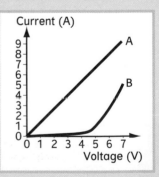

a) How, if at all, does the resistance change as the voltage is increased in:
  i) component A
  ii) component B? (2)

b) i) What is component A?
  ii) What is component B? (2)

6 This diagram shows two carbon electrodes in a beaker of copper chloride solution. The ammeter shows there is a current flowing.

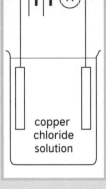

a) Copy the diagram and label
   i) the negative electrode
   ii) the positive electrode. (2)

b) Explain how the current flows
   i) in the solution
   ii) in the wires. (2)

copper chloride solution

7 a) Which magnetic pole will attract a north-seeking pole? (1)

b) A nail is attracted to a magnet. Suggest a metal from which the nail could be made. (1)

8 This diagram shows a coil of wire with a current flowing through it. A magnetic field is formed around the coil.

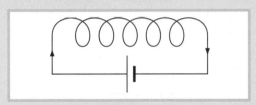

a) Copy the diagram and draw the magnetic field around the coil. (2)

b) Describe three ways of increasing the strength of the magnetic field. (3)

9 This diagram shows a relay circuit. The circuit is used to start the motor in a machine.

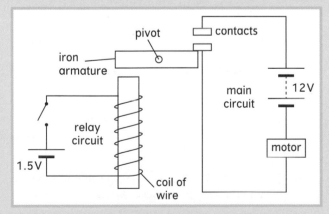

a) Explain how pressing the switch makes current flow in the motor. Include each of the labelled parts in your explanation. (5)

b) Give one other use of an electromagnet. (1)

10 This diagram shows the inside of a plug.

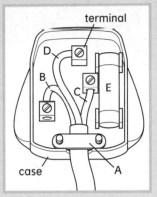

a) Name the parts of the plug labelled A to E. (5)

b) What colour are the wires B, C and D? (3)

c) What is a suitable material for
   i) the case of the plug
   ii) the terminals? (2)

d) Which wire (B, C or D) would be connected to the metal case of an appliance? (1)

e) Explain how the earth wire protects people from electric shocks. (2)

11 The following information is stamped on a kettle.

> Electrical supply 230 V 50 Hz
> Fit with a 13 A fuse

a) When the kettle is working correctly a 10 A current flows through it. What is the power of the kettle when it is working correctly? (1)

b) What is the highest current that can flow in this kettle? (1)

c) What is the purpose of the fuse? (1)

d) The kettle uses alternating current from the mains electricity supply. What is the difference between this current and the current from a cell? (1)

12 This diagram shows an electricity generator. When the coil rotates a current is induced in the wire.

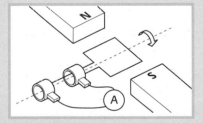

a) Name three ways to increase the size of the induced current. (3)

b) The mains supply voltage is 230 V. A telephone answering machine only needs 9 V to work properly. What component is used to reduce the mains supply voltage to a smaller voltage? (1)

Glossary

abdomen Part of the body between the chest and hip, containing organs like the stomach, liver and intestines.

absorb Soaking something up from the surroundings. A sponge absorbs water.

acid A substance that turns litmus red. Has a pH of less than 7.

acid rain Rain containing sulphuric acid and nitric acid. Acid rain has a pH of less than 5.6.

adapted When a cell or an organism has certain features to help it do a particular job. When the shape of a cell helps it to do its job it is said to be 'adapted' to its job or function.

addictive A drug that causes the user to become dependent on it.

aerobic respiration The main respiration reaction in cells. It uses oxygen from the air and glucose from food to release energy. The waste products are carbon dioxide and water.

alcohol One of a group of chemicals with similar properties. Ethanol is the scientific name for the chemical found in alcoholic drinks.

alcoholic Someone who is dependent on alcoholic drinks.

alkali A substance that turns litmus blue. An alkali has a pH of more than 7. Another name for a base that dissolves in water.

alkaline With a pH of more than 7.

alternating current An electrical current that flows one way then the other. Generators and the mains supply alternating current.

alveoli Tiny pockets in the lungs at the ends of the bronchioles, where oxygen diffuses into the blood and carbon dioxide diffuses out.

amino acids The molecules that proteins are made of.

ammeter Piece of equipment used to measure electrical current.

ampère or **amp** (**A**) The unit for measuring electrical current.

anaerobe A living thing that survives without oxygen.

anaerobic respiration Respiration without oxygen, which takes place when cells cannot get enough oxygen to meet their energy needs by aerobic respiration.

anode Positive electrode in electrolysis.

antibodies Proteins that destroy particular microbes. They are made by white blood cells.

antitoxins Chemicals that destroy toxins. They are made by white blood cells.

anus The opening at the end of the gut.

aqueous solution A solution of something in water.

armature The turning part of a motor or generator.

artery A blood vessel that carries blood away from the heart.

atom The smallest particle of an element. It has no overall electric charge.

atomic mass The relative mass of an atom. Roughly how many times heavier it is than an atom of hydrogen.

atomic number The number of protons in an element's atoms.

atrium (plural **atria**) The upper space on each side of the heart. It receives blood from the veins.

attract Pulling something closer.

auxin Plant growth hormone. It is found in the tips of shoots and roots.

bacterium (plural **bacteria**) A single-celled microbe without a cell nucleus. Bacteria can cause disease.

barrage Large dam across a river to control the flow of the tides.

base A substance that reacts with an acid to form a salt. Some bases are alkalis.

battery Two or more electrical cells used together.

bauxite Rock containing aluminium oxide.

bile An alkaline liquid made by the liver. It helps to break down fats in the small intestine and neutralises acid from the stomach.

biodegradable Something that will decay naturally.

bladder The organ that stores urine.

blast furnace A large industrial furnace for extracting iron from iron ore.

blood clot A plug formed by blood to seal a cut and stop microbes getting into the body. Often dries to form a scab.

blood vessel A tube that carries blood around the body.

boiling point When a liquid is at its boiling point it is as hot as it can get. It is evaporating as fast as it can.

breathing Using the lungs to take in and blow out air.

bronchi Pair of large air tubes from the trachea to the lungs.

bronchioles Small air tubes that branch out from the bronchi inside the lungs.

bronchitis When mucus in the lungs is infected by bacteria.

by-product A substance produced by a chemical reaction but is not needed.

calcium carbonate Compound of calcium, carbon and oxygen. The chemical name for limestone and marble.

calcium hydroxide Compound of calcium, hydrogen and oxygen. The chemical name for slaked lime.

calcium oxide Compound of calcium and oxygen. The chemical name for quicklime.

cancer When some of the body's cells divide too quickly. The lump of cells that forms (a tumour) can stop other cells doing their job.

capillaries Tiny blood vessels that link arteries and veins.

carbohydrates A group of foods used for energy e.g. starch and sugars.

carbon dioxide A colourless gas. It is produced by respiration and used up in photosynthesis.

carbon monoxide Very poisonous colourless gas. It stops red blood cells carrying enough oxygen around the body.

catalyse Speeding up a chemical reaction.

catalyst Something that speeds up a chemical reaction without being changed at the end of it.

cathode Negative electrode in electrolysis.

cell (in biology) The basic unit which living things are made of.

cell (in chemistry) Apparatus for electrolysis.

cell (in physics) The scientific name for a battery. It contains a store of chemical energy that can be turned into electrical energy.

cellulose The substance that plant cell walls are made of.

cement A mixture of powdered limestone and powdered clay which is heated in a rotary kiln.

cementing When sediments are compressed, water is squeezed out from between the particles. Some of the dissolved minerals get left behind and form crystals which stick the particles together to make rock.

chemical energy The kind of energy stored in chemicals. Food, fuels and electrical cells all contain chemical energy.

chemical formula Shorthand way of describing a substance using symbols.

chemical reaction A change that forms new substances.

chlorophyll The green substance found inside chloroplasts.

chloroplast A green disc containing chlorophyll. Found in plant cells and used to make glucose by photosynthesis.

cholesterol A fatty substance found in some foods. It can clog up arteries and lead to heart disease.

chromosome A thread-like strand found in the nucleus of a cell. Chromosomes are made of DNA and contain the 'instructions' for a living thing.

cilia Small hairs on the surface of some cells.

ciliary muscle A part of the eye. It works with the suspensory ligaments to hold the lens in place and change its shape.

ciliated epithelial cells Cells found in the lungs. Their hairs sweep dirty mucus out of the lungs.

circuit breaker An electromagnetic switch that breaks a circuit if the current gets too big.

circuit diagram Shorthand way of showing an electric circuit using symbols.

circulation system The set of organs that carry oxygen and food around the body.

coal A fossil fuel made from the remains of plants.

coke A fuel made from coal. It is almost pure carbon.

compress Making something smaller by squashing it.

concentrated A solution that contains a lot of the solute.

concentration gradient When a substance is more concentrated in one place than another.

concrete A mixture of cement, water, sand and crushed rock, which reacts together to form a hard substance used in building.

condense Turning from a gas into a liquid.

conduction The way that heat or electricity travels through solids by passing from particle to particle.

conductor A material that lets energy travel through it easily.

contract Muscle cells contract by getting shorter and fatter.

convection The way that heat travels through liquids and gases as the particles in them move about.

convection current A flow of liquid or gas caused by part of it being heated or cooled more than the rest.

core The innermost section of the Earth.

cornea The clear part of the sclera at the front of the eye. It helps to focus light on the retina.

corrode, corrosion When a substance is 'eaten' away by a chemical reaction.

crack, cracking Breaking long hydrocarbon molecules into shorter ones.

crude oil Oil as it comes out of the ground. It is mainly a mixture of hydrocarbons.

crust The rocky outer layer of the Earth.

cryolite A compound of aluminium, sodium and fluorine. It is added to bauxite to help lower the melting point when extracting aluminium.

cuticle Waterproof top surface of a leaf.

cutting A side stem taken off a plant. It is allowed to sprout roots to make a new plant.

decay When microbes eat the remains of a dead organism and it rots away.

decompose (in chemistry) Splitting up a substance by a chemical reaction.

deflect Changing something's direction.

dependent When someone feels that they have to carry on taking a drug to survive.

deposit A layer of particles that settles at the bottom of a liquid.

depressant A drug that slows down the nervous system.

diabetes When the pancreas produces too little insulin.

diabetic Someone who has diabetes.

diaphragm A sheet of muscle that separates the thorax from the abdomen. It moves to help you breathe.

diet Everything an animal, or human, eats and drinks.

diffuse, diffusion When particles mix together without anything moving them. They move from an area of where there are lots of them to where there are fewer of them.

digest, digestion Breaking down food into smaller units that the body can use. (Proteins are broken down into amino acids, carbohydrates are broken down into glucose and fats are broken down into fatty acids and glycerol.)

digestive juices Secretions that help break down foods in the digestive system.

digestive system Organ system used to break down food and change it into a form the body can use.

dilute A solution that does not contain much of the solute.

direct current An electrical current that flows in the same direction all the time. Electrical cells make direct current.

discharge Removing the electrical charge from something.

discontinuous deposition When the sedimentary rocks found at a place were not laid down continuously, but at different times.

displace Replacing one element in a compound by a more reactive one.

displacement reaction 'Competition' reaction between elements. A more reactive element displaces a less reactive one from it's compound.

DNA Complicated chemical that contains the 'instructions' for a living thing.

dormant When a volcano has not erupted for many years.

drug A chemical that changes the way the body works in some way.

earth Safety wire fitted to an appliance to stop anyone getting an electric shock by touching its metal case.

effector A part of the body that carries out a response to a stimulus (e.g. the muscle cells in the eye that open up the pupil in dim light).

efficiency The fraction of energy that is usefully transferred.

efficient When not much energy is wasted.

elastic potential energy The kind of energy stored in something that has been stretched or squashed and which can bounce back to its original shape.

electrical component Something that can be part of an electric circuit (e.g. cell, lamp).

electrical current A flow of electrical charges.

electrical energy The kind of energy carried by electricity.

electrode The place where electricity enters an electrolyte.

electrolysis Splitting up a substance by electricity.

electrolyte A liquid that conducts electricity.

electromagnet A magnet made by electricity.

electron A tiny negatively charged particle.

electroplating Using electrolysis to coat a metal object with a thin layer of another metal.

emphysema When parts of the lungs develop holes filled with liquid.

emulsification Breaking down large droplets into smaller ones to make an emulsion.

emulsion A mixture of tiny droplets of one liquid blended throughout another liquid.

energy Something that is needed to make things happen.

energy flow diagram A diagram to show energy changes.

enzyme Substance that speeds up a chemical reaction in the body (a biological catalyst). Each enzyme works best at a particular temperature and pH.

epidermis tissue Layer of cells that covers the surfaces of a leaf.

epithelium tissue Layers of cells that cover different body surfaces.

erosion When bits of weathered rock are moved by gravity, water, wind or ice.

ethanol The chemical name for the alcohol in alcoholic drinks.

ethene A hydrocarbon gas.

excretion Getting rid of the waste substances that have been made by chemical reactions in the body.

extrusive Igneous rock formed from lava cooling quickly on the Earth's surface (e.g. basalt). It has small crystals.

faeces The undigested and unabsorbed remains of food.

fats A group of foods used as a food reserve and to keep the body warm.

fault A break in rock layers.

fetus A developing baby inside a pregnant woman's body.

fibre The part of foods from plants that cannot be broken down by the body. It is the cellulose from their cell walls.

filament (in Physics) A thin wire that glows inside a light bulb.

flammable Something that burns very easily.

flower Plant organ containing smaller male and female sex organs. Fruits and seeds form from flowers.

flowering plant A plant that reproduces using seeds found in fruits.

fluid A substance which can flow (a gas or a liquid).

fold A bend in rock layers.

food poisoning When harmful microbes in food cause illness. The symptoms are often vomiting and diarrhoea.

fossil The ancient remains or traces of a plant or animal, preserved in rocks.

fossil fuel A fuel formed from the remains of plants or animals that died millions of years ago.

fraction A separated part from a mixture of liquids. It contains compounds that boil at nearly the same temperature.

fractional distillation A process for separating a mixture of liquids with different boiling points.

fractionating column The apparatus used to separate the different fractions in a mixture of liquids.

fruit An organ that carries the seeds of flowering plants. Can be fleshy or dry.

function The job that something does.

fuse A thin wire that melts and breaks if the current in a circuit gets too big.

fuse rating The maximum current that a fuse will conduct without melting.

gall bladder The organ that stores bile.

gastric juice Digestive juice made by glands in the stomach lining. It is very acidic and contains enzymes to break down proteins.

gene Part of a chromosome. It contains the 'instructions' for a particular feature (e.g. eye colour).

generator Large coil of wire with a magnet inside. When the magnet is turned, electricity is produced in the coil of wire.

gland cells Cells that make and release fluids.

glandular tissue Many gland cells grouped together.

glass A transparent solid which is made by heating limestone, sand and sodium carbonate together.

global warming Gradual heating of the Earth's atmosphere. It is caused by the 'greenhouse effect'.

glucagon A hormone that makes the amount of glucose in the blood go up.

glucose The sugar that plants make by photosynthesis, and that carbohydrates break down into in digestion.

gravitational potential energy The kind of energy stored by anything that can fall to the ground.

greenhouse effect When the Earth warms up more than it should because heat is trapped in its atmosphere and cannot escape into space.

greenhouse gas A gas that traps heat in the Earth's atmosphere (e.g. carbon dioxide).

group Column of elements with similar properties in the Periodic Table.

guard cells Cells on the underside of a leaf, which control the opening and closing of stomata.

gullet The tube from the back of the mouth to the stomach. Muscles in it contract and relax to push the food along.

heart The organ that pumps blood around the body.

heart beat One complete pumping action of the heart.

heart beat rate How many times the heart beats in a minute.

heart disease Caused by narrowing of the arteries carrying blood to the muscles of the heart; so the heart does not get enough oxygen.

heart valves Valves in the heart to stop blood flowing backwards.

heat conductor A material that lets heat energy flow through it easily.

heat energy The hotter something is, the more heat energy it has.

heat insulator A material that does not let heat energy flow through it easily.

heat radiation The waves of energy given off by something hot.

homeostasis The way the body keeps the conditions inside it (e.g. temperature) constant.

hormone A chemical 'messenger' that makes a body process happen. Hormones are secreted by glands and are carried around the body in the blood plasma.

hydrocarbon A compound made of only hydrogen and carbon.

hydroelectric power Making electricity by letting falling water (usually from a reservoir) turn turbines and generators.

hydrogen A light colourless gas. It is very flammable.

hydroxide A compound containing the elements hydrogen and oxygen, as well as another element e.g. sodium hydroxide.

hydroxide ion Ion formed when a hydroxide dissolves in water (OH-).

igneous rock Formed when hot molten magma from deep inside the Earth cools and becomes solid.

immune Protected against catching a disease because the body has already made the right antibodies.

immune system The body's ways of defending itself against disease.

immunised Protected against catching a disease by vaccination.

impulse The electrical 'message' that travels along a nerve cell.

inactive volcano One that has not erupted for many years.

indicator A dye that will change colour in acids and alkalis.

induced A current produced in a loop of wire by a changing magnetic field.

infected Having a disease caused by a microbe.

infra-red radiation Another term for heat radiation.

ingest Eating. We ingest (eat) food. Some white blood cells ingest microbes by surrounding them.

insoluble Something that will not dissolve.

insulated Covered with an electrical insulator.

insulator A material that does not let energy or electricity flow through it.

insulin A hormone that makes the amount of glucose in the blood go down.

intensity The amount of light energy.

intrusive An igneous rock formed underground from magma cooling slowly inside the Earth's crust (e.g. granite). It has large crystals.

ion An atom or group of atoms with an electrical charge.

iris The coloured part of the eye. It is a muscle that opens and closes the pupil.

iron core The iron rod placed inside an electromagnet's coil to make the electromagnet stronger.

iron ore Rock containing iron oxides.

joule (J) The unit for measuring energy.

kidneys A pair of organs used to clean the blood. They remove the urea in the blood and make it into urine.

kilowatt (kW) A unit for measuring power. 1 kW = 1000 W.

kilowatt-hour (kWh) The amount of energy transferred in an hour by an appliance.

kinetic energy The kind of energy in moving things.

lactic acid The waste product of anaerobic respiration. If it builds up in muscles it makes them ache.

large intestine The organ used to remove water from undigested food.

lava Magma that has erupted from a volcano.

leaf A plant organ that makes food using photosynthesis.

leaf mosaic Non-overlapping arrangement of a plant's leaves, so they all get sunlight for photosynthesis.

leaf stalk The part of a leaf that joins it to the stem.

lens The part of the eye which changes shape to focus light on the retina.

lifestyle Someone's way of life. It affects their health.

light energy The kind of energy given out by the Sun, light bulbs, candles, etc.

limestone A sedimentary rock made mainly of calcium carbonate.

limiting factor The factor that controls how fast a chemical reaction can go.

lithosphere A layer in the Earth made up of the crust and the top part of the mantle.

litmus A simple kind of indicator. It is red in acids and blue in alkalis.

live One of the wires that carries electricity in a plug.

liver The organ used to make and destroy substances in our bodies.

magma Hot liquid rock in the Earth's mantle or crust.

magnet Something that attracts iron.

magnetic field The space around a magnet where its effects are felt.

magnetic material Substance that is attracted to a magnet. Contains iron, nickel or cobalt.

mantle The layer of the Earth between the crust and the core.

metal A strong shiny element that can be hammered into shape. Metals are good conductors of heat and electricity.

metal hydroxide A compound of oxygen, hydrogen and a metal (e.g. sodium hydroxide).

metal oxide A compound of a metal and oxygen.

metamorphic A rock that has been changed by great heat or pressure (e.g. marble).

metre (m) A unit for measuring length.

microbe A tiny organism that can only be seen with a microscope. Microbes can cause disease.

micro-organism Another name for a microbe.

microscope Instrument used to look at very small things.

minerals (in biology) Chemical ions that are essential in small amounts for living things to stay healthy.

minerals (in geology) The chemicals that form rocks.

monomer The building blocks of polymers. Many monomers join together to make a polymer.

motor effect When a wire with an electric current flowing through it is put in a magnetic field, it moves.

mucus A thick slippery secretion.

muscle tissue Cells that can change their length and so help us to move.

National Grid System of overhead and underground cables that carry electricity around the country.

native When something occurs in nature as the element itself, not as a compound.

natural gas A fossil fuel made from the remains of animals.

naturally immune When you become immune to a disease because you have already had it.

nerve Bundle of nerve cells.

nerve cell A cell that carries messages to other nerve cells.

nerve ending A receptor in the skin which is sensitive to temperature or touch.

nervous system The organ system that carries messages around the body.

neurone Another name for a nerve cell.

neutral (in chemistry) Substance that is not an acid or an alkali. Has a pH of 7.

neutral (in physics) i) One of the wires that carries electricity in an appliance. ii) Something with equal amounts of positive and negative charge.

neutralisation The reaction between an acid and a base. A salt and water are produced.

neutralise Making something neutral.

newton (N) The unit for force.

nicotine A poisonous, addictive drug in tobacco.

non-metal Element that is not a metal.

non-renewable Resources that cannot be replaced once they have been used. Eventually they will run out.

North pole, North-seeking pole The end of a freely suspended bar magnet that points north.

nuclear accident An accident that releases radioactive material into the surroundings.

nuclear reaction A change inside atoms. A different element is produced and huge amounts of energy are released.

nutrient A substance that a living thing needs so that it can grow healthily.

ohm (W) The unit for measuring electrical resistance.

oil A fossil fuel made from the remains of animals.

optic nerve The nerve that carries messages from the retina to the brain.

ore A rock that contains useful minerals.

organ A group of different tissues working together.

organ system A collection of organs working together.

organism Any living thing. An organism must do all seven of the 'life processes'.

osmosis When water flows through a semi-permeable membrane so that the concentrations on either side become more equal.

overload A fuse is overloaded if the currents gets too big and melts it.

oxide A compound of an element and oxygen.

oxidise, oxidation Adding oxygen to a chemical, in a chemical reaction.

oxygen A colourless gas that makes up about 20% of the air. It is produced by photosynthesis and used up in respiration.

oxygen debt The amount of oxygen needed to remove the lactic acid left from anaerobic respiration.

palisade cell A cell found in leaves, which contains many chloroplasts.

palisade tissue Many palisade cells grouped together.

pancreas The organ that secretes digestive juices and hormones.

parallel (electricity) When the current in an electrical circuit can flow along different routes.

period Horizontal row of elements in the Periodic Table.

Periodic Table Chart with the chemical elements arranged in order of atomic number. Elements with similar properties appear in the same column.

permanent magnet Something that attracts iron all the time.

phloem tissue Living cells grouped together to carry dissolved food substances from the leaves to other parts of the plant.

photosynthesis Process that plants use to make their own food. It needs light to work. Carbon dioxide and water are used up. A sugar called glucose, and oxygen is produced.

photosynthese Making food by photosynthesis.

plant hormone Chemical that controls the way a plant grows.

plant organ Group of different plant tissues working together to do an important job.

plasma The liquid part of the blood, which is mainly water. It carries many substances around the body (e.g. hormones, waste carbon dioxide and urea, nutrients).

platelets Tiny pieces of cells in the blood that release chemicals to help blood to clot. They come from cells in bone marrow.

plutonium A fuel used in nuclear power stations.

pole The ends of a magnet, where its effects are strongest.

polyethene The chemical name for polythene. It is a polymer of ethene.

polymer A long molecule made from thousands of smaller ones (monomers). Plastics are polymers.

potential difference Another name for voltage.

potential energy The scientific word for 'stored' energy.

power How quickly something transfers energy.

power line Overhead or underground cables that carry electricity.

process Sorting out information.

product A substance formed by a chemical reaction.

property A way that a substance behaves.

proportional Two quantities are proportional to each other if doubling one of them makes the other one double too.

protein coat Outer coating of a virus.

proteins Important substances used for growth and repair.

proton Tiny positively charged particle in an atom's nucleus.

pupil Gap in the middle of the iris of the eye.

pure A single substance, not mixed with anything else.

pus The remains after many microbes have been ingested by white blood cells at a spot or cut.

pylon A tall tower holding up overhead power cables.

quarry A place where useful rocks are dug out of the ground.

quicklime Substance made by heating limestone. Its chemical name is calcium oxide.

radiate Giving off waves of energy. A candle radiates light and heat energy.

radiation The way heat travels as waves of energy through space or transparent materials.

radioactive waste Dangerous waste from nuclear power stations.

rate The speed of a chemical reaction.

raw material Another term for a reactant.

reactant A substance used up in a chemical reaction.

reactive A substance that is likely to react.

Reactivity Series List of metals arranged in order of how reactive they are.

receptors Cells that detect changes in the body or its surroundings.

rectum The last part of the large intestine, leading to the anus. It stores faeces.

red blood cells The cells that give blood its colour. They contain haemoglobin, which carries oxygen around the body.

reduce, reduction Taking oxygen away from a compound in a chemical reaction.

reducing agent A chemical that will reduce other substances.

refinery Place where the chemicals in crude oil are separated and purified.

reflect Bouncing something back from a surface.

reflex action An automatic response to a stimulus, often to protect the body from harm.

relax When a muscle relaxes after contracting it goes back to its original shape.

relay An electromagnetic switch for turning large currents on safely.

renewable An energy source that can be replaced or used again and again, and will never run out (e.g. solar power).

repel Pushing something away.

resistance How difficult it is for an electrical current to flow through something.

resistor An electrical component that decreases the current in a circuit.

respiration Chemical reaction inside cells to release energy from glucose.

response How the body reacts to a stimulus that has been detected (e.g. opening up the pupil in dim light).

retina The back of the eye. It contains receptors that are sensitive to light.

ripple mark Patterns left behind in sedimentary rock from the time when the sediment was under water.

rock cycle All the processes which form rocks, linked together.

root Plant organ used to hold the plant in the ground and take water and mineral salts out of the soil.

root hair cell Cell found in roots. It has a large surface area to help the cell absorb water quickly.

root hair tissue Many root hair cells grouped together.

rusting Corrosion of iron by water and oxygen.

sacrificial protection Allowing a piece of reactive metal to corrode so that an object made of a less reactive metal does not.

saliva Secretion from the salivary glands. It contains enzymes to break down starch, and mucus to help food pass smoothly down the gullet.

salt A compound formed when an acid reacts with a base.

scab Hard protective covering over a cut that forms when a blood clot dries.

sclera Protective outer layer of the eye.

secretion Useful substance (e.g. tears, saliva, hormones) made by gland cells.

sediment Tiny particles that settle to the bottom of a liquid.

sedimentary Rock formed by the compression and cementing of material that has settled at the bottom of the sea.

seed Grows into a new plant. Made by flowering plants and conifers.

semi-permeable A membrane that will let small particles, like water, through it but not large ones.

sensory neurone A nerve cell that carries messages from a receptor to the brain or spinal cord.

series Electrical components connected 'in line' so that all of an electric current flows through each one, one after another.

sex organ The stamen (male) and carpel (female) in a flower. They make the male and female sex cells.

slaked lime A base made from limestone. Its chemical name is calcium hydroxide.

small intestine The organ used to digest and absorb food.

solar cell A kind of battery that generates electricity using energy from the Sun.

soluble Something that can dissolve in a liquid (e.g. salt is soluble in water).

solution A substance dissolved in a solute.

solvent A chemical used to dissolve something.

solvent abuse Breathing in solvent fumes on purpose.

sound energy The kind of energy given out by something that makes a noise.

South Pole, South-seeking pole The end of a freely suspended bar magnet that points south.

spinal cord Large bundle of nerve cells that carries messages to and from the brain. It runs down the back, inside the spine.

stain Dye used to colour parts of a cell to make them easier to see.

stainless steel Mixture of iron with chromium, carbon and other elements. It does not rust.

starch Carbohydrate that plants use as a store of food.

stem Plant organ used to support a plant and take water and mineral salts to the leaves.

stimulus (plural **stimuli**) Change within the body or in its surroundings that a receptor detects (senses).

stomach Organ used to help break down food. It secretes digestive juices.

stomata (singular **stoma**) Small holes on the underside of leaves which let gases into and out of the leaf.

storage organ Part of a plant where food substances can be stored.

sucrose The chemical name for the sugar used in cooking. Some plants (e.g. sugar beet) make it from glucose.

sugars Group of carbohydrates that dissolve in water and taste sweet.

sulphur dioxide A gas that is produced in small quantities when fossil fuels are burnt. It is a cause of acid rain.

surface area The total area of all the surfaces of a shape.

suspensory ligaments Part of the eye. They work with the ciliary muscle to hold the lens in place and change its shape.

symbol equation Shorthand way of showing what happens in a chemical reaction using symbols.

symptom Sign that the body has a disease.

target cell A cell that is affected by a hormone.

target organ An organ that is affected by a hormone.

taste bud Receptor on the tongue which is sensitive to flavours.

tectonic plate A section of the Earth's lithosphere.

temporary magnet Something that can be made to attract iron when needed.

terminal Where an electrical connection is made.

thermal decomposition Breaking a compound down by heating it.

thermal energy Another name for heat energy.

thorax The chest, containing the lungs and heart.

tides Twice-daily rising and falling of sea level, caused by the pull of the Moon.

tissue A group of the same cells all doing the same job.

tobacco The dried leaves of the tobacco plant.

toxin Poisonous substance made by a living thing.

trachea The main air tube to the lungs.

transfer The word used for heat or energy moving from place to place.

transformer Piece of equipment that increases or decreases voltage. A step-up transformer increases voltage and a step-down transformer decreases voltage.

transmit Getting electricity from one place to another.

transpiration Loss of water from a plant's leaves.

transpiration stream The flow of water up through a plant's roots and stem to its leaves.

transport (in Earth science) When eroded fragments are moved away from their 'parent rock' by wind or water.

turbine A machine that is turned by a moving fluid.

Unit A unit for measuring the amount of electrical energy transferred. 1 Unit is the same as 1 kilowatt-hour.

universal indicator A special mixture of indicators. It gives a different colour depending on how weak or strong an acid or an alkali is.

unlike Opposite charges or magnetic poles.

uranium A fuel used in nuclear power stations.

urea Waste product from the breakdown of unwanted amino acids by the liver.

urinate Getting rid of urine when you go to the toilet.

urine Solution of the body's waste products, which are removed from the blood by the kidneys.

vaccination Being given an injection of a vaccine to help the body protect itself against a disease.

vaccine Weak or dead disease-causing microbes put into the body on purpose. White blood cells make the right antibodies and the person becomes immune.

valve Part of a vein or the heart that stops the blood in it from flowing the wrong way.

vapour Another name for a gas.

vegetable Any plant food that is not a fruit.

vein (in a plant) Bundle of tubes that carry water and food substances into and out of a leaf.

vein (in the body) Blood vessel that carries blood towards the heart.

ventilation The movement of air into and out of the lungs.

ventricle The lower space on each side of the heart. It pumps blood into arteries.

villus (plural **villi**) Projections on the lining of the small intestine. They increase its surface area and speed up absorption of nutrients into the blood.

violent A reaction that is very quick.

virus Tiny protein-coated particle that causes disease. It enters a living cell where it reproduces, damaging the cell.

viscous A liquid that is thick and 'treacly', not runny.

vitamin Chemical from a living thing that is essential in small amounts for the body to stay healthy.

volcano An opening in the Earth's crust where magma can flow out (erupt) onto the surface from time to time.

volt (**V**) The unit for measuring voltage, or potential difference.

voltage The amount of 'pushing' that an electrical cell does.

voltmeter Piece of equipment used to measure voltage.

wasted energy Energy that cannot be used.

watt (**W**) The unit for measuring power.

weathering The breaking down of rocks by rain or ice.

white blood cells Various types of cells in the blood that help to protect the body from disease.

wilt When a plant has not had enough water and goes 'floppy'.

wind turbine A kind of windmill that generates electricity using energy from the wind.

withdrawal symptom The unpleasant side-effects when someone stops taking an addictive drug.

word equation Way of showing what happens in a chemical reaction using words.

xylem cells Dead cells found in the stems and roots of plants. They are hollow.

xylem tissue Xylem cells grouped together in tubes that carry water and mineral salts up from the plant's roots to the leaves.

Index

Pearson Education
Edinburgh Gate
Harlow
Essex

ISBN 0582 43698 2

Designed and produced by Pentacor Plc, High Wycombe

Printed in Great Britain by Scotprint, Haddington

The publisher's policy is to use paper manufactured from sustainable forests.

Acknowledgments

The publisher would like to thank many people for their help, support and encouragement in the production of this book. In particular: Susan Anderson, Gary Baker, Patricia Baker, Graham Barney, John Brierley, Malcolm Burns, David Kirkby, Janet Murray, Silvia Newton, Alastair Sandiforth, and Chris Shipley: Davenant Foundation School, Loughton; Burnt Mill School, Harlow.

We are grateful to the following for permission to reproduce copyright photographs:

Ace Photo Agency pages 76 *middle*(Mauritus), 80 *top left*(Graham Young), 94 *top*(Ian Sabell), 108 *top left*(P & M Walton), 146 *top right*(PLI), 152 *bottom*(Peter Adams); Adams Picture Library page 176; Alton Towers page 178 *left*; Heather Angel/Natural Visions page 54 *middle right*; Ardea pages 62 *bottom*(P Morris), 122 *top*; Art Directors & TRIP pages 12 *bottom left*(Eric Smith), 49 *top*(M Walker), 75(H Rogers), 94 *middle left*(A Lambert), 94 *middle right*(A Lambert), 180(H Rogers), 199(H Rogers); Auto Express Picture Library page 50 *top*; Biophoto Associates pages 8 *bottom right*, 160; Gareth Boden pages 13, 17, 19, 20 *bottom*, 25, 29 *left*, 31 *bottom left*, 31 *top*, 32 *bottom*, 37, 42 *bottom*, 49 *bottom*, 55 *bottom right*, 59, 69, 70, 74 *bottom*, 76 *bottom*, 88, 93, 94 *bottom*, 96, 97, 100, 107, 110, 112 *top left*, 112 *bottom*, 117 *top*, 119, 121, 122 *bottom*, 123 *right*, 127, 128 *top*, 129 *middle*, 129 *bottom*, 135 *left*, 147 *bottom*, 151, 166 *bottom*, 169, 171, 172, 173, 175, 177 *bottom*, 178 *right*, 181 *left*, 183 *left*, 184 *bottom*, 185, 186 *bottom*, 189, 192, 200 *top*, 201 *top*, 203 *bottom*, 207 *bottom*, 208 *bottom*, 209, 212 *top*, 212 *bottom*, 213, 215, 223, 227, 228 *bottom*, 239 *bottom*, 241; Professor W J Broughton page 56 *right*; Trevor Clifford pages 92 *top*, 120 *bottom*, 166 *top*, 166 *middle*, 225; Corbis page 242; Custom Medical Stock Photo pages 80(V Zuber), 83 *bottom right*(OJ Staats); Dorling Kindersley pages 43 *top right*(Ian O'Leary), 54 *middle left*(Roger Phillips), 58 *top*(Neil Fletcher), 66 *top*(Max Alexander), 87 *top right*(Dave King), 129 *top*(Harry Taylor), 130 *top*(Colin Keates), 130 *bottom*, 136 *left*, 136 *top middle*(Colin Keates), 136 *top right*(Harry Taylor), 136 *bottom left*(Colin Keates), 136 *bottom middle*(Colin Keates), 136 *bottom right*, 201 *bottom*(Tim Ridley), 224(Susannah Price); Greg Evans International pages 12 *top right*(Greg Balfour Evans), 91 *bottom left*; Mary Evans Picture Library page 144; www.freeplay.net page 240; GeoScience Features Picture Library pages 47 *top left*, 50 *bottom*, 87 *bottom left*, 92 *bottom*, 102 *bottom right*, 131 *bottom*, 132 *left*, 143 *bottom*, 146 *top left*, 146 *bottom*, 154 *top*, 163; Robert Harding Picture Library pages 6 *bottom middle left*(Jay Thomas), 27 *top*(Dr Dennis Kunkel/Phototake NYC); C Hoseason pages 87 *bottom right*, 114 *top right*, 114 *middle left*, 117 *bottom*, 159 *top left*, 236 *bottom*, 238 *bottom*; Holt Studios International pages 64 *bottom*, 66 *middle left* & 66 *middle right*(Nigel Cattin); Hornby plc page 243; Hutchison Picture Library pages 89(Robert Francis), 147 *top*(Tony Souter); Penny Johnson page 131 above *middle*; Frank Lane Picture Agency pages 6 *top*(D Maslowski), 6 *top middle left*(Marineland), 6 *top middle right*(Silvestris), 6 *bottom right*(Sunset), 46 *bottom left*(E & D Hosking), 60(M J Thomas), 62 *top*(A Wharton), 196(Celtic Picture Library), 220(H Binz); Mark Levesley pages 43 *top left*, 46 *top*, 46 *bottom right*, 47 *bottom*, 51 *bottom*, 52 *top left*, 54 *bottom left*, 55 *top*, 55 *bottom left*, 64 *top*, 66 *bottom*, 67 *left*, 71 *right*, 80 *top right*, 87 *top left*, 90 *bottom*, 91 *top*, 103, 114 *middle*, 116, 118 *top*, 118 *bottom*, 120 *top*, 120 *bottom middle left*, 120 *bottom middle right*, 153 *bottom middle*, 158, 159 *top right*, 182 *right*, 183 *right*, 186 *top*, 200 *bottom*, 202 *top*, 216, 235, 236 *top*, 238 *top*; Miller Pattison Ltd page 174; NHPA pages 6 *bottom middle right*(Joe Blossom), 47 *top right*(Haroldo Palo Jr), 120 above *middle left*(Stephen Dalton), 132 *bottom right*, 135 *right*(N A Callow), 153 *top*(Nigel J Dennis), 153 *bottom right*(David Woodfall), 154 *bottom*(David Woodfall); Natural Science Photos page 56 *left*(C Williams); Oxford Scientific Films pages 12 *middle*(NASA), 54 *top*(Heinrich Van Den Berg/Gallo Images), 67 *right*(Chris Sharp), 120 above *middle right*(Larry Crowhurst), 126(NASA), 128 *bottom*(Andrew Plumptire), 131 *bottom middle*(Mark Hamblin), 132 *top right*(John Cancalosi), 142(Doug Allan), 195(Ronald Toms); PA News page 40; PYMCA page 184 *top*(Richard Braine); Pictor International pages 11, 42 *top*, 42 *middle*, 54 *bottom right*, 91 *bottom right*, 197 *left*, 206; Popperfoto pages 38 *left*, 74 *top*(Reuters/Domingo Giribaldi), 143 *top*(Reuters/Simon Kwong), 182 *left*(Reuters/Gary Caskey); Powerstock Zefa pages 6 *bottom left*(Kelly Mooney), 68, 77, 177 *top*; Rex Features pages 43 *bottom*, 102 *top*, 108 *bottom*, 114 *top left*, 153 *middle*, 181 *right*, 191; Coral Rogers page 32 *top*; Science Museum/Science & Society Picture Library page 86 *right*; Science Photo Library 8 *top left*(Andrew Syred), 8 *bottom left*(Secchi-Lecaque/Roussel-Uclaf/CNRI), 9(Prof P Mottia/Dept of Anatomy/University "La Sapienza", Rome), 22(NASA), 26(Juergen Berger, Max-Planck Institute), 27 *bottom*(Andrew Syred), 31 *bottom right*(BSIP VEM), 34 *top*(Mark Clarke), 34 *middle*(Dr P Marazzi), 34 *bottom left*(Dept of Medical Photography, St Stephens Hospital, London), 34 *bottom right*(Scott Camazine), 38 *right*(Dee Breger), 39 *left*(Biology Media), 39 *right*(Dr Jeremy Burgess), 51 *top*(Claude Nuridsany & Marie Perennou), 52 *top right*(Andrew Syred), 52 *bottom*(David Frazier/Agstock), 58 *middle*(J C Revy), 78 *top*(Saturn Stills), 78 *bottom*(Saturn Stills), 81(A Glauberman), 83 *bottom left*, 134 *left*(Martin Bond), 134 *right*(Sinclair Stammers), 145(Dr Ken MacDonald), 197 *right*, 208 *top*(Montreal Neuro Institute/McGill University/CNRI), 218(Peter Menzel); Skyscan Photolibrary page 48 *right*; E.F Smith Collection. Rare Books & Manuscript Library, University of Pennsylvania, Philadelphia page 86 *left*; Still Pictures page 104; The Stock Market page 102 *bottom left*; Telegraph Colour Library pages 12 *top left*(Megumi Miyake), 12 *bottom right*(F.P.G © Spencer Jones), 33(F.P.G © J Cummins), 152 *top*(David Noton), 170(I & V Krafft/HoaQui), 188(NASA), 190(V.C.L), 194(L Lefkowitz), 221(V.C.L); Topham Picturepoint pages 71 *left*, 108 *top left*, 114 *bottom*; Vitax page 80 *middle*; Simon Watts pages 15, 76 *top*, 83 *top*, 87 *middle left*, 87 *middle right*, 90 *top*, 90 *middle*, 123 *left*, 153 *bottom left*, 159 *bottom*, 202 *bottom*, 203 *top*, 207 *top*, 228 *top*, 233, 239 *top*; Wellcome Trust Medical Photographic Library pages 20 *top*(Mike Kayser), 29 *right*(National Medical Slide Bank), 48 *left*, 232; David Woodfall/Woodfall Wild Images page 131 *right*.

Cover Photos: Climber - Telegraph Colour Library(John Terence Turner)
Seedling – BSIP Marlaud/Science Photo Library
Wind farm – Russell D. Curtis/Science Photo Library